Meat, Poultry and Game
ERRATA

Page	Food No.	Correction
2	-	*Line 9 of paragraph 4* Appendix on page **138**
11	4	*Beef, extra trimmed fat, raw* Satd **33.4** not 36.2; Monounsatd **26.8** not 29.6; Polyunsatd **1.2** not 1.9
11	18	*Fillet steak, fried, lean* Satd **3.4** not 2.1; Monounsatd **2.9** not 1.8; Polyunsatd **0.9** not 0.6; Total trans **0.3** not 0.2
11	19	*Fillet steak, fried, lean and fat* Satd **3.9** not 2.7; Monounsatd **3.3** not 2.2; Polyunsatd **1.0** not 0.6; Total trans **0.4** not 0.2
15	29	*Fore-rib/rib-roast, raw, lean and fat* Satd **8.9** not 7.5; Monounsatd **8.8** not 7.4; Polyunsatd **0.7** not 0.6; Total trans **0.8** not 0.6
15	31	*Fore-rib/rib-roast, microwaved, lean and fat* Satd **9.2** not 9.3; Monounsatd **9.1** not 9.3
15	34	*Fore-rib/rib-roast, roasted, lean and fat* Monounsatd **9.1** not 9.2
23	66	*Sirloin steak, fried, lean* Satd **3.4** not 2.1; Monounsatd **3.3** not 2.1; Polyunsatd **0.7** not 0.4; Total trans **0.3** not 0.2
23	67	*Sirloin steak, fried, lean and fat* Satd **6.0** not 5.1; Monounsatd **5.9** not 5.0; Polyunsatd **1.1** not 0.6; Total trans **0.6** not 0.4
38	136	*Leg, whole, roasted medium, lean and fat* kcal **240** not 146; kJ **1008** not 602
126	-	**Loin** not BLoin
143	-	*Line 3 of paragraph 3* appropriate, are given in the **Appendices** on **pages 127** and **132**
147	-	*Table headers should read:* **Number of samples** **% loss Means and ranges**
148	-	*Line 3 of paragraph 2* (and skin) given in the **Appendices** on **pages 127** and **132**

The Royal Society of Chemistry and the Ministry of Agriculture Fisheries and Food. January 1996

Meat, Poultry and Game

Fifth supplement to the Fifth Edition of

McCance and Widdowson's

The Composition of Foods

W. Chan, J. Brown, S. M. Lee and D. H. Buss

The Royal Society of Chemistry
and
Ministry of Agriculture, Fisheries and Food

The publishers make no representation, express or implied, with regard to the accuracy of the information contained in this book and cannot accept any legal responsibility or liability for any errors or omissions that may be made.

A catalogue record for this book is available from the British Library.

ISBN 0-85186-380-9

Published by the The Royal Society of Chemistry, Cambridge, and the Ministry of Agriculture, Fisheries, and Food, London.

Photocomposed by Land and Unwin Ltd, Bugbrooke
Printed in the United Kingdom by the Bath Press, Bath

CONTENTS

ACKNOWLEDGEMENTS

Numerous people have helped during the preparation of this book.

Most of the new analyses in this book were undertaken by The Laboratory of Government Chemist under the direction of Mr I Lumley and Mrs G Holcombe. Some of the fatty acid analyses were done by Dr M A Jordan at RHM Technology.

We would like to thank the Meat and Livestock Commission and the British Poultry Meat Federation Ltd for providing additional data. The Meat and Livestock Commission and the British Chicken Information Service kindly provided the cover photographs.

We would also like to express our appreciation for all the help given to us by many people in the Ministry of Agriculture, Fisheries and Food (MAFF), The Royal Society of Chemistry (RSC) and elsewhere who were involved in the work leading up to the production of this book. In particular, we would like to thank Mrs G Tibbs (formerly RSC) for her help in the production of this supplement.

The preparation of this book was overseen by a committee which, besides the authors, comprised Dr A M Fehily (HJ Heinz Company Ltd), Ms J Higgs (Meat and Livestock Commission), Dr J M Hughes (MAFF), Miss A A Paul (MRC Dunn Nutrition Centre), Mrs P M Richardson (Northwick Park Hospital), Dr A H Skull (RSC), Professor D A T Southgate (formerly the Institute of Food Research), and Dr F J Taylor (RSC).

INTRODUCTION

This is the eighth detailed reference book on the nutrients in food, in a series replacing and extending the information in McCance and Widdowson's *The Composition of Foods*. It presents new values for the nutrients in a wide range of cuts of beef, lamb, pork, poultry, game and offals, both raw and cooked in a variety of ways, but it does not include bacon, ham and other commercial meat products, or meat-based dishes, which will be the subject of a separate book.

Because there were few new values for meats when the summary fifth edition of *The Composition of Foods* (Holland *et al.*, 1991b) was being prepared, most of the meat values in that book were taken directly from the fourth edition (Paul and Southgate, 1978) and thus reflect the composition of meat in the early 1970s. There have been substantial changes in the composition of carcase meats since then, especially reductions in the amounts of fat both on the carcase itself and after trimming in the shop or in the home. There have also been changes in cooking methods and an increase in the range of poultry products available. A wide variety of new analyses was therefore commissioned on a large number of representative samples of meat collected between 1992 and 1995, and almost all the values in this book are taken from these analyses. The number of foods included is also larger, with information on 429 items of carcase meat, poultry, game and offals compared with 142 in the fifth edition, and there is also an increase in the number of nutrients shown.

These tables are part of a continuing series produced by the Royal Society of Chemistry (RSC) and the Ministry of Agriculture, Fisheries and Food (MAFF), who have been collaborating since 1987 on the development of a comprehensive and up-to-date database on nutrients in the wide range of foods now available in Britain. The other detailed supplements in the series currently include *Cereals and Cereal Products* (Holland *et al.*,1988), *Milk Products and Eggs* (Holland *et al.*,1989), *Vegetables, Herbs and Spices* (Holland *et al.*, 1991a), *Fruit and Nuts* (Holland *et al.*, 1992a), *Vegetable Dishes* (Holland *et al.*, 1992b), *Fish and Fish Products* (Holland *et al.*, 1993), and *Miscellaneous Foods* (Chan *et al.*, 1994). Computerised versions are also available, details of which can be obtained from the Royal Society of Chemistry.

Methods

The selection of foods and the determination of nutrient values follows the general principles used for previous books in this series. The types and cuts of meat, poultry, game and offals were chosen to represent as far as possible those most widely available in Britain, and the scientific literature was then reviewed for information on their composition. Most earlier values could not be used because of the changes in composition; and because values from other countries in general refer to meat which is differently cut and trimmed, almost all the information in this book has been derived from new analyses.

1

Literature values

Literature information was only included where full details of the samples were known and where they were clearly similar to the meats now available in British shops or caterers. Suitable methods of analysis also had to be used, and the results had to be available in sufficient detail for a full assessment.

Analyses

A large number of new analyses were needed for the production of these tables, and these were commissioned by the Ministry of Agriculture, Fisheries and Food from the Laboratory of the Government Chemist (LGC) and from the Meat and Livestock Commission (MLC). Some fatty acid analyses were also carried out by RHM Technology. As it was not practical to analyse all of the large number of types of meat, both raw and cooked in a variety of ways, for every nutrient, the following protocols were followed.

LGC programme: For all the analyses carried out by the LGC, up to 10 samples of each item were bought from a wide variety of supermarkets, butchers and market stalls in South-East England, Bristol, Leeds and Edinburgh. A major source of variation in the composition of beef, lamb and pork is the proportion of lean and fat, which can vary substantially at the retail level, so care was taken to ensure that the samples were as representative as possible in terms of fatness.

Each of the pieces of meat that was to be analysed raw was first separated where possible into its lean, fat, and any inedible material (bone and gristle) by trimming with a sharp kitchen knife in the way that consumers might do, and each of these parts was weighed. For poultry, representative samples of which were also bought from a variety of outlets, any skin was removed, and the light and dark parts of the meat as well as the bone were separated and each was weighed. Up to 10 additional samples of each type of meat were bought for each cooking method that was specified, and after these had been cooked at the LGC (for details, see the Appendix on page 000) any juices were discarded and these samples too were separated where appropriate into their lean and fat, inedible parts, light meat, dark meat and skin, and each was weighed. No attempt was made in this programme to dissect out all the intermuscular fat from within the lean, and the method also meant that in some cuts small amounts of fat could on occasion remain attached to the outside of the lean or small amounts of lean remain attached to the fat.

As in previous supplements, the individual samples of each type of meat were combined before analysis, but the lean and fat, and the light meat, dark meat and skin were always kept separate from each other. Protein (N × 6.25), fat and water were determined in the lean from every raw and cooked item, in most separated samples of fat, but skin from different parts of the poultry was combined before analysis. Nutrient values for whole cuts (i.e. lean and fat together, or meat and skin together) were then calculated from the nutrients in the separated parts and the measured weights of these parts. In each case, values for raw lean and fat were used for raw samples, and values for cooked lean and fat were used for the cooked samples. A wide range of minerals and vitamins was also analysed in selected samples, both raw and cooked. For the samples that were not analysed for all the minerals and vitamins, values from similar cuts and cooking methods were interpolated, usually in proportion to the protein content of the samples.

The analytical methods were as described in the fifth edition of *The Composition of Foods* (Holland *et al.*, 1991b). Individual fatty acids were determined as their methyl esters by capillary gas chromatography, and cholecalciferol and 25-hydroxy cholecalciferol were determined in selected samples of lean and fat by quantitative HPLC. The amounts of vitamin D in the main tables are the sum of the cholecalciferol and 5 times the amount of the more active 25-hydroxy cholecalciferol, but since there is as yet no generally accepted factor for the vitamin D activity of dietary 25-hydroxy cholecalciferol, the analysed amount of each form in the main types of meat is given in an Appendix. Further details of each determination can be provided on request.

MLC programme: In separate studies, the MLC determined the amounts of protein (N × 6.25), fat and water in a large number of beef, lamb and pork cuts taken from representative animals of known breed, age, gender and conformation. In each case, as much intermuscular and subcutaneous fat as possible was carefully separated from the lean with a sharp knife, and all lean and connective tissue separated from the fat, before the samples were analysed. These values were for raw meat only, and are given separately in the tables as extra trimmed lean and fat. Additional values for selected cuts are presented in the Appendix on page 127.

Arrangement of the tables

Food groups

For ease of reference, the values have been brought together in the following eight groups: Beef and veal; lamb (separated into British and New Zealand); pork; chicken; turkey; other poultry; game; and offals. For beef, lamb and pork, the cuts are listed alphabetically, and the first set of values for each is for the raw meat followed by values for the items as cooked. For most items, values are given for the lean only, and then for the cut with the amount of fat associated with it when sold. A number of values are in addition presented per 100g of meat with its bone, for convenience of use in those dietary surveys where the weight of meat including the bone has been recorded, and values for some larger joints are also presented for the weight including bone for the convenience of caterers. For those who wish to recalculate the amounts of nutrients in meats with different (measured) proportions of lean and fat, average values are also given at the beginning of the sections on beef, lamb and pork for the nutrients in trimmed and in extra-trimmed lean and fat.

The sequence is different within the chicken and turkey groups, where all the values for raw meat are presented first, followed by values for casseroled and stewed poultry, then fried and grilled, and finally roasted birds. Within these groups, values for the separated dark meat, light meat and skin are usually given first, followed by the main parts in alphabetical order, and then the whole plucked and dressed birds (excluding giblets).

Numbering system

As in previous supplements, the foods have been numbered in sequence together with a unique two digit prefix. For this supplement, the prefix is 18, so that the full code numbers for the first and last foods in this book are 18-001 and 18-429, and these are the numbers that will be used in nutrient databank applications.

Description and number of samples

The names of the foods indicate whether fat, skin or bone was included in the item, and the cooking method. Information is then given on the number of samples taken for analysis or, where the values were calculated from the nutrients in the separated lean, fat, dark meat, light meat or skin, the measured proportions of each part are shown.

Nutrients

The nutrient values for each food are shown on four consecutive pages as in most previous books in this series. The presentation follows the established pattern, with the nutrients on three of the pages being similar in all supplements, while those on the second page are the ones most appropriate to (in this supplement) meats and meat products. All values are given per 100 grams of the food as described.

Proximates: – The first page for each food shows the edible portion of the food as described, and then the amounts (in grams per 100g) of water, total nitrogen, protein, fat, and available carbohydrate expressed as its monosaccharide equivalent. The food's energy value is also given in kilocalories and in kilojoules. Protein was derived from the nitrogen values by multiplying them by 6.25 after subtracting any non-protein nitrogen, and the energy values were derived by multiplying the amounts of protein and fat (and carbohydrate, if present) by the factors in **Table 1**.

Table 1: - Energy conversion factors

	kcal/g	kJ/g
Protein	4	17
Fat	9	37
Available carbohydrate expressed as monosaccharide	3.75	16

Carbohydrates, fibre and fats: – The second page gives the amounts of saturated fatty acids, total monounsaturates (i.e. *cis* and *trans* together) and total polyunsaturates (also *cis* and *trans* together) in each meat, plus the amounts of cholesterol. There is an additional column showing the total amounts of *trans* fatty acids in each food, but these cannot be added to the values in the previous columns since this would double-count the *trans* acids. Selected values for the main individual saturated, monounsaturated and polyunsaturated fatty acids are given in the Appendix on page 148, but the amounts of a much wider range of fatty acids in the full range of foods will be published in a forthcoming supplement. The second page also includes columns for starch, sugars and dietary fibre, but because meats other than offals do not contain measurable amounts of any of these, most values have been imputed as zero. This layout has been used for consistency with the next supplement on the composition of meat products and meat dishes, many of which include carbohydrates.

Minerals and vitamins: – The range of minerals and vitamins shown is the same as in previous books. The values for total carotene and vitamin E have been corrected for the relative activities of the different fractions using the factors given in the fifth edition of *The Composition of Foods* (Holland *et al.*, 1991b), but because the amounts of these fractions are so small they are not given in this supplement; they are, however, available on request. Retinol and carotene values below the limit of detection (usually 5μg per 100g) are given as trace. The vitamin D activity of the food has been taken as the sum of the cholecalciferol and five times the amount of any of the more active metabolite 25-hydroxycholecalciferol known to be present. Except in the offals, vitamin C was not measured but imputed to be zero.

Appendices

This supplement contains a number of appendices. The first indicates the parts of the carcase from which the main items are derived and some of the more common names for different cuts of meat in different parts of Britain. The second shows for raw and cooked beef, lamb and pork the means and ranges of fat, lean and inedible matter (bone and sometimes gristle) in the individual samples prior to pooling. Then for poultry, the means and ranges of white meat, dark meat, skin and bone are shown. The next appendix gives the composition of some new leaner cuts of meat produced by seam butchery, and this is followed by appendixes describing the cooking methods used and the means and ranges found for the loss of weight (fat and water) when different cuts of meat were cooked. Three further appendixes give details of the individual fatty acids in the major types of meat, the amounts of vitamin A in offals and of vitamin D in selected meats.

Nutrient variability

Almost all foods vary somewhat in composition, and this is especially true for meat. First, there are substantial variations in the amount and composition of both the lean and the fat with breed, age, season and the proportion of grass and concentrates in the animal's feed. The amount of fat in the lean also varies with the original fatness of the animal, and the amount of fat in the fat (and its fatty acid composition) varies with the cut. There can also be substantial and variable trimming of the fat before the meat is sold, and in the home and in catering establishments before or after cooking. Furthermore, the length and temperature of cooking, and the size of the meat being cooked, all influence the nutrient content of the meat as eaten. The amount of fat in meat has been steadily decreasing and is continuing to decrease, so to help allow for all possible options the data in the main tables gives the nutrients in lean and fat separately as well as in the main cuts of meat. In addition, the nutrients in fully trimmed meat where *all* the visible fat has been removed are shown.

The values for most foods in this book are averages based on a number of representative items or cuts bought between 1992 and 1995, but it is important to remember that besides becoming leaner, the nutritional value of any meat can at any time be significantly different from the average. Although some indication of variability is given in the Appendixes, it is important when using these tables to ensure that the product is as similar as possible to that described here. This requires that, whenever the contribution of meats to nutrient intakes is being assessed, the amount of visible fat remaining on the meat and the method and extent of cooking are as far as possible taken into account.

It is also important to bear in mind that for some related cuts, apparent (but usually small) differences in composition may reflect analytical variations as much as real differences in composition.

The introduction to the fifth edition of *The Composition of Foods* contains a more detailed description of these and many other factors that should be taken into account in the proper use of food composition tables. Users of the present supplement are advised to read them and take them to heart.

References to introductory text

Paul, A.A. and Southgate, D.A.T. (1978) *McCance and Widdowson's The Composition of Foods*, 4th edition, Her Majesty's Stationery Office, London

Holland, B., Unwin, I.D., and Buss, D.H. (1988) *Cereals and Cereal Products.* Third supplement to *McCance and Widdowson's The Composition of Foods*, Royal Society of Chemistry, Cambridge

Holland, B., Unwin, I.D., and Buss, D.H. (1989) *Milk Products and Eggs.* Fourth supplement to *McCance and Widdowson's The Composition of Foods*, Royal Society of Chemistry, Cambridge

Holland, B., Unwin, I.D., and Buss, D.H. (1991a) *Vegetables, Herbs and Spices.* Fifth supplement to *McCance and Widdowson's The Composition of Foods*, Royal Society of Chemistry, Cambridge

Holland, B., Welch, A.A., Unwin, I.D., Buss, D.H., Paul, A.A. and Southgate, D.A.T. (1991b) *McCance and Widdowson's The Composition of Foods*, 5th edition, Royal Society of Chemistry, Cambridge

Holland, B., Unwin, I.D., and Buss, D.H. (1992a) *Fruit and Nuts.* First supplement to 5th edition of *McCance and Widdowson's The Composition of Foods*. Royal Society of Chemistry, Cambridge

Holland, B., Welch, A.A., and Buss, D.H. (1992b) *Vegetable Dishes.* Second supplement to 5th edition of *McCance and Widdowson's The Composition of Foods*. Royal Society of Chemistry, Cambridge

Holland, B., Brown, J., and Buss, D.H. (1993) *Fish and Fish Products.* Third supplement to 5th edition of *McCance and Widdowson's The Composition of Foods*. Royal Society of Chemistry, Cambridge

Chan, W., Brown, J., and Buss, D.H. (1994) *Miscellaneous Foods.* Fourth supplement to 5th edition of *McCance and Widdowson's The Composition of Foods*. Royal Society of Chemistry, Cambridge

The
Tables

Symbols and abbreviations used in the tables

Symbols

0	None of the nutrient is present
Tr	Trace
N	The nutrient is present in significant quantities but there is no reliable information on the amount
( )	Estimated value

Abbreviations

MLC	Meat and Livestock Commission
LGC	Laboratory of The Government Chemist
Satd	Saturated
Monounsatd	Monounsaturated
Polyunsatd	Polyunsaturated
Trypt	Tryptophan

No. 18-	Food	Description and main data sources	Edible Proportion	Water g	Total Nitrogen g	Protein g	Fat g	Carbohydrate g	Energy value kcal	kJ
1	**Beef**, average, trimmed lean, *raw*	LGC; average of 10 different cuts	1.00	71.9	3.60	22.5	5.1	0	136	571
2	extra trimmed lean, *raw*	MLC; weighted average of 13 different cuts	1.00	72.5	3.45	21.6	5.1	0	132	556
3	trimmed fat, *raw*	LGC; average of 10 different cuts	1.00	35.0	3.02	18.9	53.6	0	558	2305
4	extra trimmed fat, *raw*	MLC; weighted average of 13 different cuts	1.00	20.9	1.08	6.8	72.2	0	677	2786
5	fat, *cooked*	LGC; average of 8 different cuts	1.00	33.6	2.48	15.5	52.3	0	533	2199
6	**Braising steak**, *raw*, lean	10 samples	1.00	72.1	3.49	21.8	5.7	0	139	582
7	–, lean and fat	Calculated from 92% lean and 7% fat	0.99	69.4	3.31	20.7	8.6	0	160	670
8	*braised*, lean	10 samples	1.00	55.5	5.50	34.4	9.7	0	225	944
9	–, lean and fat	Calculated from 90% lean and 9% fat	1.00	53.1	5.26	32.9	12.7	0	246	1029
10	*slow cooked*, lean	10 samples	1.00	60.0	5.02	31.4	7.9	0	197	826
11	–, lean and fat	Calculated from 88% lean and 10% fat	1.00	57.1	4.64	29.0	11.2	0	217	907
12	**Brisket**, *raw*, lean	10 samples	1.00	71.0	3.38	21.1	6.1	0	139	584
13	–, lean and fat	Calculated from 79% lean and 19% fat	0.99	61.8	2.94	18.4	16.0	0	218	905
14	*boiled*, lean	10 samples	1.00	57.2	5.02	31.4	11.0	0	225	941
15	–, lean and fat	Calculated from 82% lean and 15% fat	0.98	51.7	4.45	27.8	17.4	0	268	1116
16	**Fillet steak**, *raw*, lean	17 samples	1.00	72.3	3.39	21.2	6.1	0	140	586
17	–, lean and fat	Calculated from 94% lean and 6% fat	1.00	70.7	3.34	20.9	7.9	0	155	648
18	*fried*, lean	16 samples	1.00	62.6	4.51	28.2	7.9	0	184	772
19	–, lean and fat	Calculated from 96% lean and 4% fat	1.00	61.9	4.48	28.0	8.9	0	192	805
20	*grilled*, lean	20 samples	1.00	62.6	4.66	29.1	8.0	0	188	791
21	–, lean and fat	Calculated from 97% lean and 3% fat	1.00	61.6	4.59	28.7	9.5	0	200	839

No. 18-	Food	Starch g	Total sugars g	Dietary fibre Southgate method g	Dietary fibre Englyst method g	Fatty acids Satd g	Fatty acids cis & trans Mono-unsatd g	Fatty acids cis & trans Poly-unsatd g	Total trans g	Cholesterol mg
1	**Beef**, average, trimmed lean, *raw*	0	0	0	0	2.2	2.3	0.3	0.1	58
2	extra trimmed lean, *raw*	0	0	0	0	2.0	2.0	0.2	N	N
3	trimmed fat, *raw*	0	0	0	0	24.9	24.2	1.7	2.3	72
4	extra trimmed fat, *raw*	0	0	0	0	36.2	29.6	1.9	N	N
5	fat, *cooked*	0	0	0	0	24.3	23.4	1.8	2.4	97
6	**Braising steak**, *raw*, lean	0	0	0	0	2.4	2.5	0.3	0.2	63
7	-, lean and fat	0	0	0	0	3.8	3.8	0.4	0.3	62
8	*braised*, lean	0	0	0	0	4.1	4.1	0.6	0.3	100
9	-, lean and fat	0	0	0	0	5.3	5.2	0.8	0.5	100
10	*slow cooked*, lean	0	0	0	0	3.4	3.3	0.5	0.3	90
11	-, lean and fat	0	0	0	0	4.8	4.7	0.7	0.4	87
12	**Brisket**, *raw*, lean	0	0	0	0	2.5	2.5	0.2	0.2	54
13	-, lean and fat	0	0	0	0	6.7	7.1	0.6	0.7	53
14	*boiled*, lean	0	0	0	0	4.6	5.0	0.4	0.5	77
15	-, lean and fat	0	0	0	0	7.2	7.7	0.7	0.7	76
16	**Fillet steak**, *raw*, lean	0	0	0	0	2.8	2.4	0.3	0.2	61
17	-, lean and fat	0	0	0	0	3.8	3.1	0.4	0.2	65
18	*fried*, lean	0	0	0	0	2.1	1.8	0.6	0.2	81
19	-, lean and fat	0	0	0	0	2.7	2.2	0.6	0.2	83
20	*grilled*, lean	0	0	0	0	3.6	3.2	0.5	0.2	71
21	-, lean and fat	0	0	0	0	4.4	3.9	0.5	0.2	72

Inorganic constituents per 100g food

No. 18-	Food	Na	K	Ca	Mg	P	Fe (mg)	Cu	Zn	Cl	Mn	Se (µg)	I
1	**Beef**, average, trimmed lean, *raw*	63	350	5	22	200	1.8	0.03	4.1	51	0.01	7	10
2	extra trimmed lean, *raw*	N	N	N	N	N	N	N	N	N	N	N	N
3	trimmed fat, *raw*	26	140	5	9	79	0.7	0.02	1.1	28	Tr	2	10
4	extra trimmed fat, *raw*	N	N	N	N	N	N	N	N	N	N	N	N
5	fat, *cooked*	35	200	6	12	110	1.0	0.01	1.5	39	0.01	3	14
6	**Braising steak**, *raw*, lean	64	320	5	20	190	1.5	Tr	6.0	54	Tr	7	17
7	-, lean and fat	60	300	5	19	180	1.4	Tr	5.6	51	Tr	7	16
8	*braised*, lean	62	340	8	23	220	2.7	Tr	9.5	62	Tr	11	15
9	-, lean and fat	60	330	8	22	210	2.6	Tr	8.7	61	Tr	10	15
10	*slow cooked*, lean	53	260	7	20	200	2.5	Tr	8.5	31	Tr	10	12
11	-, lean and fat	50	250	7	19	190	2.3	Tr	7.6	31	Tr	9	12
12	**Brisket**, *raw*, lean	59	330	4	21	190	1.7	0.03	3.4	49	Tr	7	10
13	-, lean and fat	50	280	4	18	160	1.5	0.03	2.9	43	Tr	6	9
14	*boiled*, lean	50	250	9	21	200	2.2	Tr	6.6	56	0.02	10	10
15	-, lean and fat	46	230	8	19	170	1.9	Tr	5.6	51	0.02	9	10
16	**Fillet steak**, *raw*, lean	44	340	4	23	210	2.1	Tr	2.8	51	0.02	7	7
17	-, lean and fat	43	330	4	22	200	2.0	Tr	2.7	50	0.02	7	7
18	*fried*, lean	68	390	6	27	240	2.3	Tr	5.1	56	Tr	9	10
19	-, lean and fat	67	390	6	27	230	2.3	Tr	5.0	56	Tr	9	10
20	*grilled*, lean	70	400	6	27	250	2.3	N	5.2	58	N	10	11
21	-, lean and fat	67	390	6	27	240	2.3	N	5.1	57	N	10	11

No. 18-	Food	Retinol µg	Carotene µg	Vitamin D µg	Vitamin E mg	Thiamin mg	Ribo- flavin mg	Niacin mg	Trypt 60 mg	Vitamin B_6 mg	Vitamin B_{12} µg	Folate µg	Panto- thenate mg	Biotin µg	Vitamin C mg
1	**Beef**, average, trimmed lean, *raw*	Tr	Tr	0.5	0.13	0.10	0.21	5.0	4.7	0.53	2	19	0.75	1	0
2	extra trimmed lean, *raw*	N	N	N	N	N	N	N	N	N	N	N	N	N	0
3	trimmed fat, *raw*	Tr	Tr	Tr	0.06	0.04	0.13	1.2	1.7	0.17	1	18	0.43	1	0
4	extra trimmed fat, *raw*	N	N	N	N	N	N	N	N	N	N	N	N	N	0
5	fat, *cooked*	Tr	Tr	Tr	0.08	0.05	0.18	1.6	1.8	0.23	2	26	0.60	2	0
6	**Braising steak**, *raw*, lean	Tr	Tr	0.5	0.09	0.07	0.28	4.2	4.6	0.45	2	50	0.61	1	0
7	-, lean and fat	Tr	Tr	0.5	0.09	0.07	0.27	3.9	4.3	0.42	2	53	0.59	1	0
8	*braised*, lean	Tr	8	0.8	0.02	0.05	0.26	5.2	8.0	0.34	3	54	0.55	2	0
9	-, lean and fat	Tr	7	0.7	0.03	0.05	0.26	4.9	7.5	0.33	3	52	0.57	2	0
10	*slow cooked*, lean	Tr	8	0.7	0.04	0.04	0.20	3.5	7.0	0.28	3	49	0.40	2	0
11	-, lean and fat	Tr	7	0.5	0.04	0.04	0.19	3.2	6.3	0.27	2	45	0.40	2	0
12	**Brisket**, *raw*, lean	Tr	Tr	0.5	0.14	0.09	0.24	4.9	4.4	0.51	2	14	0.71	1	0
13	-, lean and fat	Tr	Tr	0.4	0.12	0.08	0.21	4.0	3.7	0.43	2	14	0.63	1	0
14	*boiled*, lean	Tr	8	0.7	0.01	0.04	0.22	4.3	7.0	0.19	3	14	0.50	2	0
15	-, lean and fat	Tr	7	0.6	0.02	0.04	0.20	3.7	6.0	0.19	2	15	0.49	2	0
16	**Fillet steak**, *raw*, lean	Tr	Tr	0.5	0.12	0.14	0.28	4.7	4.5	0.47	2	10	1.05	1	0
17	-, lean and fat	Tr	Tr	0.4	0.12	0.13	0.28	4.6	4.3	0.46	2	11	1.02	1	0
18	*fried*, lean	Tr	8	0.6	N	0.12	0.25	6.2	6.3	0.59	2	16	0.83	2	0
19	-, lean and fat	Tr	8	0.6	N	0.12	0.25	6.1	6.2	0.58	2	17	0.83	2	0
20	*grilled*, lean	Tr	8	0.7	0.06	0.12	0.26	6.4	6.5	0.61	2	17	0.85	2	0
21	-, lean and fat	Tr	8	0.6	0.06	0.12	0.26	6.3	6.4	0.60	2	17	0.84	2	0

Composition of food per 100g

No. 18-	Food	Description and main data sources	Edible Proportion	Water g	Total Nitrogen g	Protein g	Fat g	Carbohydrate g	Energy value kcal	kJ
22	**Fillet steak**, from steakhouse, lean	7 samples	1.00	64.0	4.59	28.7	7.0	0	178	747
23	-, lean and fat	Calculated from 85% lean and 14% fat	1.00	63.5	4.56	28.5	7.8	0	184	773
24	**Flank**, *raw*, lean	6 samples	1.00	68.8	3.63	22.7	9.3	0	175	730
25	-, lean and fat	Calculated from 76% lean and 23% fat	1.00	60.2	3.15	19.7	20.8	0	266	1105
26	*pot-roasted*, lean	6 samples	1.00	53.7	5.09	31.8	14.0	0	253	1059
27	-, lean and fat	Calculated from 76% lean and 21% fat	0.98	47.4	4.34	27.1	22.3	0	309	1286
28	**Fore-rib/rib-roast**, *raw*, lean	10 samples	1.00	71.7	3.44	21.5	6.5	0	145	606
29	-, lean and fat	Calculated from 75% lean and 25% fat	1.00	61.4	3.01	18.8	19.8	0	253	1052
30	*microwaved*, lean	10 samples	1.00	52.0	5.60	35.0	11.4	0	243	1017
31	-, lean and fat	Calculated from 78% lean and 22% fat	1.00	46.8	4.86	30.4	20.5	0	306	1275
32	-, -, weighed with bone	Calculated from no. 31	0.83	38.8	4.03	25.2	17.0	0	254	1057
33	*roasted*, lean	10 samples	1.00	55.9	5.33	33.3	11.4	0	236	988
34	-, lean and fat	Calculated from 78% lean and 22% fat	1.00	49.8	4.64	29.1	20.4	0	300	1250
35	-, -, weighed with bone	Calculated from no. 34	0.84	41.8	3.90	24.4	17.1	0	252	1048
36	**Mince**, *raw*	10 samples	1.00	62.0[a]	3.15	19.7	16.2[b]	0	225	934
37	*microwaved*	10 samples	1.00	55.3	4.22	26.4	17.5	0	263	1096
38	*stewed*	10 samples	1.00	64.4	3.49	21.8	13.5	0	209	870
39	*frozen, stewed*	10 samples	1.00	65.5[c]	3.09	19.3	14.1[d]	0	204	850
40	extra lean, *raw*	10 samples	1.00	68.1[e]	3.50	21.9	9.6[f]	0	174	728
41	-, *stewed*	17 samples	1.00	66.6	3.95	24.7	8.7	0	177	742

[a] Water ranged from 57.3g to 70.0g per 100g
[c] Water ranged from 58.2g to 74.3g per 100g
[e] Water ranged from 63.8g to 72.6g per 100g

[b] Fat ranged from 7.8g to 26.5g per 100g
[d] Fat ranged from 3.6g to 23.9g per 100g
[f] Fat ranged from 3.9g to 16.9g per 100g

Beef *continued*

No. Food 18-	Starch g	Total sugars g	Dietary fibre Southgate method g	Englyst method g	Fatty acids cis & trans Satd g	Mono-unsatd g	Poly-unsatd g	Total trans g	Cholesterol mg
22 **Fillet steak**, from steakhouse, lean	0	0	0	0	3.2	2.8	0.4	0.2	79
23 -, lean and fat	0	0	0	0	3.6	3.1	0.4	0.2	81
24 **Flank**, *raw*, lean	0	0	0	0	3.8	4.3	0.4	0.3	58
25 -, lean and fat	0	0	0	0	8.7	9.6	0.8	0.8	60
26 *pot-roasted*, lean	0	0	0	0	5.7	6.5	0.6	0.5	91
27 -, lean and fat	0	0	0	0	9.1	10.1	0.9	0.9	88
28 **Fore-rib/rib-roast**, *raw*, lean	0	0	0	0	2.9	2.8	0.2	0.2	56
29 -, lean and fat	0	0	0	0	7.5	7.4	0.6	0.6	59
30 *microwaved*, lean	0	0	0	0	5.0	4.9	0.4	0.4	85
31 -, lean and fat	0	0	0	0	9.3	9.3	0.7	0.8	87
32 -, -, weighed with bone	0	0	0	0	7.7	7.7	0.6	0.7	72
33 *roasted*, lean	0	0	0	0	5.1	5.0	0.4	0.4	81
34 -, lean and fat	0	0	0	0	9.2	9.2	0.7	0.8	83
35 -, -, weighed with bone	0	0	0	0	7.7	7.7	0.6	0.7	70
36 **Mince**, *raw*	0	0	0	0	7.1	7.1	0.5	0.7	60
37 *microwaved*	0	0	0	0	7.6	7.7	0.7	0.8	80
38 *stewed*	0	0	0	0	5.9	5.9	0.5	0.6	79
39 *frozen, stewed*	0	0	0	0	6.2	6.2	0.6	0.7	58
40 extra lean, *raw*	0	0	0	0	4.2	4.1	0.4	0.4	56
41 -, *stewed*	0	0	0	0	3.8	3.8	0.3	0.4	75

Inorganic constituents per 100g food

No. 18-	Food	Na	K	Ca	Mg	P	Fe	Cu	Zn	Cl	Mn	Se	I
						mg						µg	
22	**Fillet steak**, from steakhouse, lean	61	410	5	32	250	3.6	0.10	4.5	64	0.02	9	11
23	-, lean and fat	61	410	5	32	250	3.5	0.10	4.4	64	0.02	9	11
24	**Flank**, *raw*, lean	64	350	5	23	210	1.8	0.03	3.6	52	Tr	7	10
25	-, lean and fat	54	300	5	19	170	1.5	0.03	3.0	46	Tr	6	10
26	*pot-roasted*, lean	51	250	7	20	190	2.1	Tr	6.9	53	0.02	10	10
27	-, lean and fat	45	230	7	18	160	1.8	Tr	5.5	48	0.02	8	10
28	**Fore-rib/rib-roast**, *raw*, lean	61	340	5	22	200	1.7	0.03	3.5	50	Tr	7	10
29	-, lean and fat	52	290	5	19	170	1.5	0.03	2.9	44	Tr	6	10
30	*microwaved*, lean	60	290	8	23	220	2.8	0.04	6.3	61	0.01	12	13
31	-, lean and fat	54	270	7	21	190	2.4	0.04	5.2	55	0.01	7	13
32	-, -, weighed with bone	45	220	6	17	160	2.0	0.03	4.3	46	0.01	8	11
33	*roasted*, lean	57	360	9	23	220	2.0	Tr	7.4	64	Tr	11	12
34	-, lean and fat	54	320	8	20	190	1.8	Tr	6.1	58	Tr	10	12
35	-, -, weighed with bone	43	270	7	17	160	1.5	Tr	5.1	49	Tr	8	10
36	**Mince**, *raw*	80	260	9	17	160	1.4	Tr	3.9	76	Tr	7	9
37	*microwaved*	91	290	12	20	190	2.0	Tr	5.2	110	0.02	9	16
38	*stewed*	73	210	20	15	150	2.2	0.10	5.0	63	0.02	7	14
39	*frozen, stewed*	62	210	17	17	140	2.4	0.12	3.7	78	0.02	6	10
40	*extra lean, raw*	90	290	10	19	180	1.5	0.06	4.4	85	Tr	7	11
41	*-, stewed*	75	280	14	18	170	2.3	0.08	5.6	61	Tr	8	10

No. 18-	Food	Retinol µg	Carotene µg	Vitamin D µg	Vitamin E mg	Thiamin mg	Ribo-flavin mg	Niacin mg	Trypt 60 mg	Vitamin B6 mg	Vitamin B12 µg	Folate µg	Panto-thenate mg	Biotin µg	Vitamin C mg
22	**Fillet steak**, from steakhouse, lean	Tr	8	0.7	0.08	0.14	0.31	5.7	7.1	0.81	2	8	1.28	2	0
23	-, lean and fat	Tr	8	0.6	0.08	0.14	0.31	5.6	6.9	0.80	2	9	1.27	2	0
24	**Flank**, *raw*, lean	Tr	Tr	0.5	0.15	0.10	0.25	5.3	4.8	0.54	2	16	0.77	1	0
25	-, lean and fat	Tr	Tr	0.4	0.13	0.09	0.22	4.3	3.9	0.45	2	16	0.68	1	0
26	*pot-roasted*, lean	Tr	8	0.7	0.04	0.04	0.19	3.0	7.1	0.23	3	16	0.48	2	0
27	-, lean and fat	Tr	6	0.5	0.06	0.04	0.18	2.6	6.1	0.22	2	17	0.48	2	0
28	**Fore-rib/rib-roast**, *raw*, lean	Tr	8	0.5	0.14	0.10	0.24	5.1	4.6	0.52	2	15	0.74	1	0
29	-, lean and fat	Tr	6	0.4	0.12	0.09	0.21	4.1	5.3	0.43	2	16	0.66	1	0
30	*microwaved*, lean	Tr	8	0.8	0.01	0.05	0.26	4.8	7.9	0.34	3	14	0.49	2	0
31	-, lean and fat	Tr	6	0.6	0.02	0.05	0.24	4.1	6.4	0.31	3	16	0.51	2	0
32	-, -, weighed with bone	Tr	5	0.5	0.02	0.04	0.20	3.4	5.3	0.26	2	13	0.42	2	0
33	*roasted*, lean	Tr	8	0.8	0.22	0.06	0.16	5.3	7.5	0.40	3	18	0.53	2	0
34	-, lean and fat	Tr	6	0.6	0.19	0.06	0.17	4.5	6.1	0.36	3	19	0.54	2	0
35	-, -, weighed with bone	Tr	5	0.5	0.16	0.05	0.14	3.8	5.1	0.30	2	16	0.45	2	0
36	**Mince**, *raw*	Tr	Tr	0.5	0.17	0.06	0.13	5.8	3.6	0.37	2	14	0.49	1	0
37	*microwaved*	Tr	8	0.6	0.31	0.08	0.31	8.0	4.3	0.38	3	17	0.53	2	0
38	*stewed*	9	25	0.9	0.34	0.03	0.19	4.6	4.4	0.28	2	17	0.36	5	0
39	*frozen, stewed*	Tr	8	0.4	0.31	0.05	0.19	3.0	3.5	0.13	2	30	0.41	1	0
40	*extra lean, raw*	Tr	Tr	0.5	0.18	0.07	0.15	6.4	4.0	0.42	2	16	0.55	1	0
41	-, *stewed*	Tr	8	0.6	0.30	0.03	0.13	4.8	4.5	0.16	3	20	0.36	2	0

Beef *continued*

Composition of food per 100g

No. 18-	Food	Description and main data sources	Edible Proportion	Water g	Total Nitrogen g	Protein g	Fat g	Carbohydrate g	Energy value kcal	Energy value kJ
42	**Mince patties**, *barbecued*	10 samples	1.00	54.5	4.74	29.6	16.2	0	264	1103
43	**Rump steak**, *raw*, lean	10 samples	1.00	72.7	3.52	22.0	4.1	0	125	526
44	-, lean and fat	Calculated from 88% lean and 11% fat	1.00	68.2	3.31	20.7	10.1	0	174	726
45	*barbecued*, lean	10 samples	1.00	62.4	4.99	31.2	5.7	0	176	741
46	-, lean and fat	Calculated from 91% lean and 8% fat	1.00	59.3	4.72	29.5	9.4	0	203	849
47	*fried*, lean	10 samples	1.00	61.7	4.94	30.9	6.6	0	183	770
48	-, lean and fat	Calculated from 87% lean and 12% fat	1.00	57.2	4.54	28.4	12.7	0	228	953
49	*grilled*, lean	10 samples	1.00	62.9	4.96	31.0	5.9	0	177	745
50	from steakhouse, lean	10 samples	1.00	63.0	4.77	29.8	4.7	0	162	681
51	-, lean and fat	Calculated from 85% lean and 14% fat	1.00	57.9	4.40	27.5	11.4	0	213	889
52	strips, *stir-fried*, lean	10 samples	1.00	57.9	5.17	32.3	8.8	0	208	875
53	-, lean and fat	Calculated from 86% lean and 13% fat	1.00	53.8	4.75	29.7	14.4	0	248	1038
54	**Silverside**, *raw*, lean	10 samples	1.00	72.2	3.81	23.8	4.3	0	134	564
55	-, lean and fat	Calculated from 80% lean and 18% fat	0.99	62.5	3.26	20.4	14.8	0	215	894
56	*pot-roasted*, lean	10 samples	1.00	58.9	5.44	34.0	6.3	0	193	811
57	-, lean and fat	Calculated from 85% lean and 14% fat	1.00	54.0	4.96	31.0	13.7	0	247	1034
58	*salted*, *raw*, lean	Calculated from 80% lean and 19% fat	1.00	72.0	3.07	19.2	7.0	0	140	585
59	-, lean and fat	Calculated from 80% lean and 19% fat	1.00	63.1	2.61	16.3	18.0	0	227	943
60	*salted*, *boiled*, lean	Calculated from 88% lean and 11% fat	1.00	60.4	4.86	30.4	6.9	0	184	772
61	-, lean and fat	Calculated from 88% lean and 11% fat	1.00	56.5	4.46	27.9	12.5	0	224	937

No. Food 18-	Starch g	Total sugars g	Dietary fibre Southgate method g	Dietary fibre Englyst method g	Fatty acids Satd g	Fatty acids cis & trans Mono-unsatd g	Fatty acids cis & trans Poly-unsatd g	Total trans g	Cholesterol mg
42 **Mince patties**, *barbecued*	0	0	0	0	7.1	7.1	0.6	0.7	90
43 **Rump steak**, *raw*, lean	0	0	0	0	1.7	1.7	0.3	0.1	59
44 -, lean and fat	0	0	0	0	4.3	4.4	0.6	0.3	60
45 *barbecued*, lean	0	0	0	0	2.4	2.4	0.4	0.1	76
46 -, lean and fat	0	0	0	0	4.0	4.1	0.6	0.2	76
47 *fried*, lean	0	0	0	0	2.5	2.6	0.9	0.1	86
48 -, lean and fat	0	0	0	0	4.9	5.2	1.6	0.3	84
49 *grilled*, lean	0	0	0	0	2.5	2.5	0.5	0.1	76
50 from steakhouse, lean	0	0	0	0	2.0	2.0	0.3	0.1	73
51 -, lean and fat	0	0	0	0	4.9	4.9	0.7	0.3	74
52 strips, *stir-fried*, lean	0	0	0	0	3.3	3.5	1.2	0.2	92
53 -, lean and fat	0	0	0	0	5.6	5.9	1.8	0.4	91
54 **Silverside**, *raw*, lean	0	0	0	0	1.6	2.0	0.2	0.1	61
55 -, lean	0	0	0	0	5.9	6.9	0.7	0.5	57
56 *pot-roasted*, lean	0	0	0	0	2.5	2.9	0.3	0.1	93
57 -, lean and fat	0	0	0	0	5.6	6.3	0.7	0.3	92
58 salted, *raw*, lean	0	0	0	0	2.5	3.5	0.3	0.2	49
59 -, lean and fat	0	0	0	0	6.8	8.9	0.8	0.6	45
60 salted, *boiled*, lean	0	0	0	0	2.5	3.4	0.3	0.2	74
61 -, lean and fat	0	0	0	0	4.7	6.2	0.6	0.4	73

Beef continued

Inorganic constituents per 100g food

No. 18-	Food	Na	K	Ca	Mg	P	Fe	Cu	Zn	Cl	Mn	Se	I
							mg					µg	
42	**Mince patties**, *barbecued*	98	320	20	23	220	3.0	0.06	6.2	98	0.02	10	15
43	**Rump steak**, *raw*, lean	60	370	4	23	210	2.1	0.04	3.8	39	Tr	7	11
44	-, lean and fat	56	350	4	22	200	2.0	0.04	3.5	38	Tr	7	11
45	*barbecued*, lean	78	460	8	29	270	3.2	0.10	5.1	61	0.04	10	11
46	-, lean and fat	74	430	8	27	250	3.0	0.09	4.8	58	0.04	9	11
47	*fried*, lean	78	390	5	25	240	3.0	0.02	5.2	50	0.02	10	9
48	-, lean and fat	71	360	5	23	220	2.7	0.02	4.7	47	0.02	9	9
49	*grilled*, lean	74	430	7	29	260	2.5	0.04	5.6	62	0.02	10	12
50	from steakhouse, lean	72	410	7	28	250	2.4	0.04	5.4	60	0.02	10	11
51	-, lean and fat	66	370	7	25	230	2.2	0.04	4.8	56	0.02	9	11
52	strips, *stir-fried*, lean	78	450	7	30	270	2.6	0.04	5.8	64	0.02	11	12
53	-, lean and fat	71	410	7	27	250	2.4	0.04	5.2	60	0.02	10	12
54	**Silverside**, *raw*, lean	62	350	6	23	200	2.0	0.02	3.8	52	Tr	8	6
55	-, lean and fat	53	300	5	19	170	1.7	0.02	3.2	45	Tr	7	6
56	*pot-roasted*, lean	58	320	5	23	220	2.8	0.08	5.2	45	0.02	11	16
57	-, lean and fat	54	290	5	21	200	2.5	0.07	4.6	44	0.02	10	16
58	salted, *raw*, lean	640	120	4	19	180	1.5	0.03	3.1	1070	Tr	6	9
59	-, lean and fat	550	98	6	17	150	1.3	0.03	2.6	920	Tr	5	8
60	salted, *boiled*, lean	1020	190	10	17	150	2.0	Tr	5.3	1690	0.02	10	11
61	-, lean and fat	940	170	12	16	130	1.8	Tr	4.8	1570	0.02	9	11

No. Food 18-	Retinol µg	Carotene µg	Vitamin D µg	Vitamin E mg	Thiamin mg	Ribo-flavin mg	Niacin mg	Trypt 60 mg	Vitamin B6 mg	Vitamin B12 µg	Folate µg	Panto-thenate mg	Biotin µg	Vitamin C mg
42 **Mince patties**, *barbecued*	Tr	8	0.7	0.09	0.06	0.26	8.7	5.4	0.42	3	18	0.72	2	0
43 **Rump steak**, *raw*, lean	Tr	Tr	0.5	0.05	0.10	0.25	5.4	4.9	0.66	2	16	0.67	1	0
44 -, lean and fat	Tr	Tr	0.4	0.04	0.09	0.23	4.9	4.5	0.61	2	17	0.65	1	0
45 *barbecued*, lean	Tr	8	0.7	0.20	0.15	0.32	6.8	7.0	0.36	3	10	0.78	2	0
46 -, lean and fat	Tr	7	0.6	0.19	0.14	0.31	6.3	6.5	0.34	2	11	0.75	2	0
47 *fried*, lean	Tr	8	0.7	0.18	0.14	0.29	5.9	6.7	0.63	2	18	0.74	2	0
48 -, lean and fat	Tr	7	0.6	N	0.13	0.27	5.3	6.0	0.57	2	16	0.70	2	0
49 *grilled*, lean	Tr	8	0.7	0.07	0.13	0.28	6.8	7.0	0.65	3	18	0.91	2	0
50 from steakhouse, lean	Tr	8	0.7	N	0.13	0.27	6.5	6.7	0.63	2	17	0.88	2	0
51 -, lean and fat	Tr	7	0.6	N	0.12	0.25	5.7	5.9	0.57	2	18	0.83	2	0
52 strips, *stir-fried*, lean	Tr	8	0.7	0.06	0.21	0.30	6.8	7.2	0.73	3	5	0.94	2	0
53 -, lean and fat	Tr	7	0.6	N	0.19	0.28	6.1	6.4	0.66	2	7	0.88	2	0
54 **Silverside**, *raw*, lean	Tr	Tr	0.5	0.15	0.10	0.27	5.6	5.0	0.57	2	16	0.81	1	0
55 -, lean and fat	Tr	Tr	0.4	0.13	0.09	0.23	4.6	4.1	0.48	2	15	0.70	1	0
56 *pot-roasted*, lean	Tr	8	0.8	0.05	0.17	0.26	5.0	7.4	0.33	2	7	0.43	2	0
57 -, lean and fat	Tr	7	0.6	0.05	0.15	0.24	4.5	6.5	0.31	2	9	0.45	2	0
58 salted, *raw*, lean	Tr	Tr	0.4	0.12	0.08	0.21	4.5	4.0	0.46	2	13	0.65	1	0
59 -, lean and fat	Tr	Tr	0.3	0.10	0.07	0.18	3.7	3.3	0.38	2	12	0.56	1	0
60 salted, *boiled*, lean	Tr	8	0.7	0.10	0.05	0.27	2.6	6.8	0.39	2	12	0.40	2	0
61 -, lean and fat	Tr	7	0.6	0.09	0.05	0.25	2.4	6.1	0.36	2	12	0.40	2	0

Beef continued

Composition of food per 100g

No. 18-	Food	Description and main data sources	Edible Proportion	Water g	Total Nitrogen g	Protein g	Fat g	Carbohydrate g	Energy value kcal	kJ
62	**Sirloin joint**, *roasted*, lean	10 samples	1.00	59.2	5.18	32.4	6.5	0	188	791
63	-, lean and fat	Calculated from 82% lean and 16% fat	1.00	54.5	4.77	29.8	12.6	0	233	973
64	**Sirloin steak**, *raw*, lean	10 samples	1.00	71.5	3.76	23.5	4.5	0	135	566
65	-, lean and fat	Calculated from 80% lean and 18% fat	1.00	64.5	3.46	21.6	12.7	0	201	837
66	*fried*, lean	10 samples	1.00	61.8	4.61	28.8	8.2	0	189	793
67	-, lean and fat	Calculated from 82% lean and 16% fat	0.99	56.6	4.28	26.8	14.0	0	233	974
68	*grilled rare*, lean	10 samples	1.00	67.2	4.22	26.4	6.7	0	166	697
69	-, lean and fat	Calculated from 83% lean and 16% fat	1.00	61.7	4.02	25.1	12.8	0	216	900
70	*grilled medium-rare*, lean	19 samples	1.00	63.9	4.26	26.6	7.7	0	176	737
71	-, lean and fat	Calculated from 86% lean and 12% fat	0.99	58.9	3.97	24.8	12.6	0	213	888
72	*grilled well-done*, lean	10 samples	1.00	55.7	5.42	33.9	9.9	0	225	943
73	-, lean and fat	Calculated from 86% lean and 13% fat	0.99	52.7	5.09	31.8	14.4	0	257	1073
74	from steakhouse, lean	7 samples	1.00	62.5	5.04	31.5	5.1	0	172	724
75	-, lean and fat	Calculated from 85% lean and 14% fat	1.00	58.3	4.74	29.6	10.7	0	215	899
76	**Stewing steak**, *raw*, lean	10 samples	1.00	73.4	3.62	22.6	3.5	0	122	514
77	-, lean and fat	Calculated from 90% lean and 9% fat	1.00	70.1	3.54	22.1	6.4	0	146	613
78	*pressure cooked*, lean	10 samples	1.00	58.2	5.62	35.1	6.5	0	199	837
79	-, lean and fat	Calculated from 89% lean and 10% fat	1.00	56.3	5.44	34.0	9.0	0	217	911
80	*stewed*, lean	10 samples	1.00	61.6	5.12	32.0	6.3	0	185	777
81	-, lean and fat	Calculated from 84% lean and 14% fat	1.00	59.4	4.67	29.2	9.6	0	203	852

22

Beef *continued*

No. 18-	Food	Starch g	Total sugars g	Dietary fibre Southgate method g	Dietary fibre Englyst method g	Fatty acids Satd g	Fatty acids cis & trans Mono-unsatd g	Fatty acids cis & trans Poly-unsatd g	Total trans g	Cholesterol mg
62	**Sirloin joint,** *roasted,* lean	0	0	0	0	2.9	2.7	0.2	0.2	79
63	-, lean and fat	0	0	0	0	5.6	5.2	0.4	0.4	85
64	**Sirloin steak,** *raw,* lean	0	0	0	0	2.0	1.9	0.2	0.1	51
65	-, lean and fat	0	0	0	0	5.6	5.5	0.5	0.4	57
66	*fried,* lean	0	0	0	0	2.1	2.1	0.4	0.2	82
67	-, lean and fat	0	0	0	0	5.1	5.0	0.6	0.4	89
68	*grilled rare,* lean	0	0	0	0	3.0	2.9	0.2	0.2	64
69	-, lean and fat	0	0	0	0	5.8	5.6	0.4	0.4	74
70	*grilled medium-rare,* lean	0	0	0	0	3.4	3.3	0.3	0.2	67
71	-, lean and fat	0	0	0	0	5.6	5.4	0.4	0.4	70
72	*grilled well-done,* lean	0	0	0	0	4.4	4.3	0.3	0.3	83
73	-, lean and fat	0	0	0	0	6.5	6.3	0.5	0.4	87
74	from steakhouse, lean	0	0	0	0	2.2	2.2	0.3	0.1	77
75	-, lean and fat	0	0	0	0	4.6	4.7	0.5	0.3	83
76	**Stewing steak,** *raw,* lean	0	0	0	0	1.4	1.6	0.2	0.1	67
77	-, lean and fat	0	0	0	0	2.6	2.9	0.4	0.2	69
78	*pressure cooked,* lean	0	0	0	0	2.4	2.7	0.8	0.1	100
79	-, lean and fat	0	0	0	0	3.4	3.8	1.1	0.2	105
80	*stewed,* lean	0	0	0	0	2.3	2.6	0.8	0.1	91
81	-, lean and fat	0	0	0	0	3.7	4.2	0.9	0.2	91

Inorganic constituents per 100g food

No. 18-	Food	Na	K	Ca	Mg	P	Fe	Cu	Zn	Cl	Mn	Se	I
		mg										µg	
62	**Sirloin joint, roasted,** lean	55	350	7	24	220	2.0	Tr	5.7	60	Tr	11	15
63	roasted, lean and fat	53	330	7	22	200	1.9	Tr	5.0	57	Tr	10	15
64	**Sirloin steak, raw,** lean	70	360	5	22	200	1.6	0.02	4.0	49	Tr	8	11
65	-, lean and fat	62	320	5	20	180	1.5	0.02	3.5	46	Tr	7	11
66	fried, lean	69	400	6	27	240	2.3	0.04	5.2	58	0.02	10	11
67	-, lean and fat	64	370	6	25	220	2.1	0.03	4.6	56	0.02	9	12
68	grilled rare, lean	63	370	6	25	220	2.1	0.04	4.8	53	0.02	9	10
69	-, lean and fat	60	340	6	23	210	2.0	0.03	4.3	52	0.02	8	11
70	grilled medium-rare, lean	65	370	7	24	220	1.4	Tr	4.3	57	Tr	10	13
71	-, lean and fat	60	340	7	22	200	1.3	Tr	3.9	54	Tr	9	13
72	grilled well-done, lean	81	470	7	32	290	2.7	0.04	6.1	68	0.02	11	13
73	-, lean and fat	76	440	7	30	260	2.5	0.04	5.5	65	0.02	10	14
74	from steakhouse, lean	76	440	7	30	270	2.5	0.04	5.7	63	0.02	10	12
75	-, lean and fat	71	410	7	28	250	2.3	0.04	5.1	61	0.02	9	13
76	**Stewing steak, raw,** lean	69	360	5	21	190	2.1	0.04	5.7	69	Tr	7	12
77	-, lean and fat	66	340	5	20	180	2.0	0.04	5.3	66	Tr	7	13
78	pressure cooked, lean	60	290	19	23	220	2.8	0.04	9.5	35	0.01	12	13
79	-, lean and fat	60	290	18	23	220	2.7	0.04	8.7	38	0.01	11	14
80	stewed, lean	54	270	17	21	200	2.6	0.04	8.6	32	0.01	11	12
81	-, lean and fat	51	250	15	19	180	2.3	0.04	7.5	33	0.01	10	12

No. 18-	Food	Retinol µg	Carotene µg	Vitamin D µg	Vitamin E mg	Thiamin mg	Riboflavin mg	Niacin mg	Trypt 60 mg	Vitamin B6 mg	Vitamin B12 µg	Folate µg	Pantothenate mg	Biotin µg	Vitamin C mg
62	**Sirloin joint**, *roasted*, lean	Tr	8	0.7	0.02	0.13	0.28	4.7	7.3	0.58	3	7	0.41	2	0
63	-, *roasted*, lean and fat	Tr	7	0.6	0.03	0.12	0.27	4.2	6.3	0.52	3	11	0.45	2	0
64	**Sirloin steak**, *raw*, lean	Tr	Tr	0.5	0.15	0.10	0.26	5.5	4.9	0.56	2	16	0.80	1	0
65	-, lean and fat	Tr	Tr	0.4	0.14	0.09	0.24	4.7	4.2	0.49	2	17	0.75	1	0
66	*fried*, lean	Tr	8	0.7	N	0.12	0.26	6.3	6.5	0.60	2	17	0.85	2	0
67	-, lean and fat	Tr	7	0.5	N	0.11	0.25	5.5	5.6	0.54	2	19	0.82	2	0
68	*grilled rare*, lean	Tr	8	0.6	0.06	0.11	0.24	5.8	5.9	0.55	2	15	0.78	2	0
69	-, lean and fat	Tr	7	0.5	0.07	0.10	0.24	5.2	5.2	0.50	2	18	0.77	2	0
70	*grilled medium-rare*, lean	Tr	12	0.5	0.02	0.07	0.16	6.9	5.7	0.58	2	14	0.64	2	0
71	-, lean and fat	Tr	10	0.4	0.03	0.07	0.16	6.1	5.1	0.53	2	15	0.62	2	0
72	*grilled well-done*, lean	Tr	8	0.8	0.08	0.15	0.31	7.4	7.6	0.71	3	20	1.00	2	0
73	-, lean and fat	Tr	7	0.7	0.08	0.14	0.30	6.7	6.8	0.65	3	21	0.96	2	0
74	from steakhouse, lean	Tr	8	0.7	N	0.13	0.28	6.9	7.1	0.66	3	18	0.93	2	0
75	-, lean and fat	Tr	7	0.6	N	0.12	0.27	6.2	6.3	0.60	2	20	0.90	2	0
76	**Stewing steak**, *raw*, lean	Tr	Tr	0.8	0.20	0.07	0.27	4.2	4.5	0.45	2	5	0.65	1	0
77	-, lean and fat	Tr	Tr	0.7	0.20	0.07	0.26	4.0	4.3	0.42	2	6	0.65	1	0
78	*pressure cooked*, lean	Tr	8	0.8	0.07	0.03	0.19	3.3	7.9	0.28	3	N	0.42	2	0
79	-, lean and fat	Tr	7	0.7	0.08	0.04	0.20	3.2	7.3	0.29	3	N	0.48	2	0
80	*stewed*, lean	Tr	8	0.7	0.19	0.02	0.15	2.6	7.2	0.23	3	8	0.31	2	0
81	-, lean and fat	Tr	7	0.6	0.17	0.03	0.15	2.4	6.2	0.23	2	11	0.30	2	0

Beef and veal

18-082 to 18-095

Composition of food per 100g

No. Food 18-	Description and main data sources	Edible Proportion	Water g	Total Nitrogen g	Protein g	Fat g	Carbo-hydrate g	Energy value kcal	kJ
82 **Stewing steak**, *frozen, stewed, lean*	10 samples	1.00	61.4	4.75	29.7	8.5	0	195	819
83 -, lean and fat	Calculated from 85% lean and 13% fat	0.99	58.9	4.30	26.9	11.6	0	212	887
84 **Topside**, *raw, lean*	10 samples	1.00	72.8	3.68	23.0	2.7	0	116	491
85 -, lean and fat	Calculated from 84% lean and 15% fat	1.00	65.8	3.26	20.4	12.9	0	198	824
86 *microwaved, lean*	10 samples	1.00	60.6	5.60	35.0	4.6	0	181	765
87 -, lean and fat	Calculated from 89% lean and 11% fat	1.00	56.8	5.14	32.1	11.1	0	228	956
88 *roasted medium-rare, lean*	10 samples	1.00	62.2	5.15	32.2	5.1	0	175	736
89 -, lean and fat	Calculated from 87% lean and 12% fat	1.00	57.6	4.78	29.9	11.4	0	222	930
90 *roasted well-done, lean*	10 samples	1.00	56.9	5.79	36.2	6.3	0	202	849
91 -, lean and fat	Calculated from 88% lean and 11% fat	1.00	53.0	5.25	32.8	12.5	0	244	1020
Veal									
92 **Escalope**, *raw*	9 samples	1.00	75.1	3.63	22.7	1.7	0	106	449
93 *fried*	9 samples	1.00	58.7	5.39	33.7	6.8	0	196	825
94 **Mince**, *raw*	5 samples	1.00	70.1[a]	3.25	20.3	7.0[b]	0	144	604
95 *stewed*	5 samples	1.00	63.7	4.21	26.3	11.1	0	205	858

[a] Water ranged from 68.6g to 75.8g per 100g

[b] Fat ranged from 2.6g to 10.5g per 100g

No. 18-	Food	Starch g	Total sugars g	Dietary fibre		Fatty acids cis & trans			Total trans g	Cholesterol mg
				Southgate method g	Englyst method g	Satd g	Mono-unsatd g	Poly-unsatd g		
82	**Stewing steak**, frozen, *stewed*, lean	0	0	0	0	3.1	3.6	1.1	0.2	85
83	-, lean and fat	0	0	0	0	4.3	4.9	1.3	0.3	82
84	**Topside**, *raw*, lean	0	0	0	0	1.1	1.2	0.2	0.1	50
85	-, lean and fat	0	0	0	0	5.4	5.8	0.8	0.3	48
86	*microwaved*, lean	0	0	0	0	1.9	2.1	0.2	0.1	85
87	-, lean and fat	0	0	0	0	4.7	5.1	0.5	0.4	82
88	*roasted medium-rare*, lean	0	0	0	0	2.1	2.3	0.2	0.1	68
89	-, lean and fat	0	0	0	0	4.8	5.2	0.5	0.4	71
90	*roasted well-done*, lean	0	0	0	0	2.6	2.8	0.3	0.2	88
91	-, lean and fat	0	0	0	0	5.2	5.7	0.6	0.4	83
Veal										
92	**Escalope**, *raw*	0	0	0	0	0.6	0.7	0.3	Tr	52
93	*fried*	0	0	0	0	1.8	2.6	1.9	0.1	110
94	**Mince**, *raw*	0	0	0	0	2.9	3.0	0.5	0.3	62
95	*stewed*	0	0	0	0	4.7	4.8	0.8	0.4	80

Beef and veal

Inorganic constituents per 100g food

No. 18-	Food	Na	K	Ca	Mg	P	Fe	Cu	Zn	Cl	Mn	Se	I
						mg						µg	
82	**Stewing steak**, frozen, *stewed*, lean	47	220	19	21	170	3.0	0.12	7.7	51	Tr	10	8
83	-, lean and fat	44	210	17	19	150	2.7	0.10	6.7	48	Tr	9	8
84	**Topside**, *raw*, lean	77	390	6	25	220	1.9	0.08	4.0	49	0.02	8	10
85	-, lean and fat	67	340	5	22	190	1.7	0.07	3.5	44	0.02	7	9
86	*microwaved*, lean	60	290	8	23	220	2.8	0.04	6.3	35	0.01	12	13
87	-, lean and fat	55	270	7	21	200	2.6	0.04	5.7	33	0.01	11	12
88	*roasted medium-rare*, lean	66	390	5	25	230	2.5	0.07	5.6	56	0.02	10	9
89	-, lean and fat	62	360	5	23	210	2.3	0.06	5.1	54	0.02	9	10
90	*roasted well-done*, lean	62	410	8	27	230	2.9	0.04	6.5	36	0.01	12	13
91	-, lean and fat	57	370	7	24	210	2.6	0.04	5.8	34	0.01	11	12
Veal													
92	**Escalope**, *raw*	59	350	4	24	230	0.6	Tr	2.4	54	0.02	9	9
93	*fried*	86	460	6	32	300	0.9	Tr	3.1	77	0.02	11	8
94	**Mince**, *raw*	83	260	9	17	170	0.8	Tr	3.0	79	0.02	7	10
95	*stewed*	74	290	16	21	210	1.1	Tr	3.9	75	0.02	9	8

No. 18-	Food	Retinol µg	Carotene µg	Vitamin D µg	Vitamin E mg	Thiamin mg	Ribo-flavin mg	Niacin mg	Trypt 60 mg	Vitamin B6 mg	Vitamin B12 µg	Folate µg	Panto-thenate mg	Biotin µg	Vitamin C mg
82	**Stewing steak**, frozen, *stewed*, lean	Tr	8	0.7	0.08	0.03	0.21	3.4	6.7	0.15	2	12	0.72	2	0
83	-, lean and fat	Tr	7	0.6	0.08	0.03	0.19	3.1	5.8	0.15	2	13	0.67	2	0
84	**Topside**, *raw*, lean	Tr	Tr	0.5	0.15	0.09	0.19	5.3	5.0	0.55	2	24	0.66	1	0
85	-, lean and fat	Tr	Tr	0.4	0.13	0.08	0.17	4.6	4.3	0.48	2	22	0.59	1	0
86	*microwaved*, lean	Tr	8	0.8	0.05	0.05	0.30	4.4	7.9	0.42	3	67	0.49	2	0
87	-, lean and fat	Tr	7	0.7	0.05	0.05	0.28	4.0	7.1	0.39	3	61	0.47	2	0
88	*roasted medium-rare*, lean	Tr	Tr	0.4	0.04	0.06	0.32	5.4	6.6	0.54	2	14	0.55	2	0
89	-, lean and fat	Tr	Tr	0.4	0.05	0.06	0.30	4.9	5.9	0.50	2	15	0.55	2	0
90	*roasted well-done*, lean	Tr	8	0.8	0.08	0.09	0.29	5.8	8.1	0.56	3	21	0.60	2	0
91	-, lean and fat	Tr	7	0.7	0.08	0.08	0.27	5.2	7.2	0.51	3	20	0.56	2	0
	Veal														
92	**Escalope**, *raw*	Tr	Tr	1.3	0.26	0.12	0.23	7.8	4.8	0.65	2	23	0.87	1	0
93	*fried*	6	Tr	1.4	0.39	0.08	0.25	7.8	7.6	0.70	4	17	1.02	5	0
94	**Mince**, *raw*	Tr	Tr	1.2	0.17	0.06	0.14	5.9	3.7	0.49	2	14	0.51	1	0
95	*stewed*	Tr	8	1.5	0.16	0.23	0.18	4.9	4.8	0.30	3	15	0.56	2	0

Lamb

18-096 to 18-114

Composition of food per 100g

No. 18-	Food	Description and main data sources	Edible Proportion	Water g	Total Nitrogen g	Protein g	Fat g	Carbo-hydrate g	Energy value kcal	kJ
96	**Lamb**, average, trimmed lean, *raw*	LGC; average of 8 different cuts	1.00	70.6	3.23	20.2	8.3	0	156	651
97	extra trimmed lean, raw	MLC; weighted average of 7 different cuts	1.00	71.6	3.21	20.0	7.5	0	148	618
98	trimmed fat, *raw*	LGC; average of 8 different cuts	1.00	34.7	2.13	13.3	51.6	0	518	2135
99	extra trimmed fat, *raw*	MLC; weighted average of 7 different cuts	1.00	15.4	1.04	6.5	77.2	0	721	2967
100	fat, cooked	LGC; average of 8 different cuts	1.00	28.3	2.64	15.4	56.3	0	568	2345
101	**Best end neck cutlets**, *raw*, lean and fat	Calculated from 66% lean and 34% fat	1.00	53.9	2.61	16.3	27.9	0	316	1309
102	-, -, weighed with bone	Calculated from no. 101	0.74	39.3	1.91	11.9	20.4	0	231	957
103	*barbecued*, lean	10 samples	1.00	58.5	4.43	27.7	13.9	0	236	985
104	-, -, weighed with fat and bone	Calculated from no. 103	0.51	29.8	2.26	14.1	7.1	0	120	502
105	-, lean and fat	Calculated from 73% lean and 27% fat	1.00	48.6	3.88	24.3	27.2	0	342	1420
106	-, -, weighed with bone	Calculated from no. 105	0.72	35.0	2.80	17.5	19.6	0	246	1023
107	*grilled*, lean	33 samples	1.00	57.4	4.56	28.5	13.8	0	238	995
108	-, -, weighed with fat and bone	Calculated from no. 107	0.48	27.6	2.19	13.7	6.6	0	114	477
109	-, lean and fat	Calculated from 68% lean and 32% fat	1.00	46.1	3.91	24.5	29.9	0	367	1523
110	-, -, weighed with bone	Calculated from no. 109	0.72	33.2	2.82	17.6	21.0	0	259	1076
111	**Breast**, *raw*, lean	10 samples	1.00	66.0	3.14	19.6	11.2	0	179	748
112	-, lean and fat	Calculated from 66% lean and 33% fat	1.00	54.8	2.59	16.2	24.7	0	287	1189
113	*roasted*, lean	10 samples	1.00	54.4	4.27	26.7	18.5	0	273	1138
114	-, lean and fat	Calculated from 62% lean and 36% fat	1.00	45.5	3.59	22.4	29.9	0	359	1487

No. 18-	Food	Starch g	Total sugars g	Dietary fibre Southgate method g	Englyst method g	Fatty acids Satd g	cis & trans Mono-unsatd g	Poly-unsatd g	Total trans g	Cholesterol mg
96	**Lamb**, average, trimmed lean, *raw*	0	0	0	0	3.8	3.2	0.4	0.6	74
97	extra trimmed lean, raw	0	0	0	0	N	N	N	N	N
98	trimmed fat, *raw*	0	0	0	0	26.3	19.5	2.3	4.8	92
99	extra trimmed fat, *raw*	0	0	0	0	N	N	N	N	N
100	fat, cooked	0	0	0	0	28.4	21.6	2.4	5.2	100
101	**Best end neck cutlets**, *raw*, lean and fat	0	0	0	0	13.6	10.4	1.4	2.2	76
102	-, -, weighed with bone	0	0	0	0	9.9	7.6	1.0	1.6	56
103	*barbecued*, lean	0	0	0	0	6.5	5.2	0.7	1.1	98
104	-, -, weighed with fat and bone	0	0	0	0	3.3	2.6	0.4	0.5	50
105	-, lean and fat	0	0	0	0	13.2	10.2	1.4	2.2	98
106	-, -, weighed with bone	0	0	0	0	9.5	7.1	1.0	1.6	71
107	*grilled*, lean	0	0	0	0	6.5	5.1	0.7	1.0	100
108	-, -, weighed with fat and bone	0	0	0	0	3.1	2.4	0.3	0.5	48
109	-, lean and fat	0	0	0	0	14.5	11.2	1.5	2.4	105
110	-, -, weighed with bone	0	0	0	0	10.4	8.1	1.1	1.7	76
111	**Breast**, *raw*, lean	0	0	0	0	5.2	4.2	0.5	0.9	76
112	-, lean and fat	0	0	0	0	11.9	9.4	1.2	2.2	80
113	*roasted*, lean	0	0	0	0	8.6	7.0	0.9	1.5	95
114	-, lean and fat	0	0	0	0	14.3	11.4	1.4	2.7	93

Inorganic constituents per 100g food

No. 18-	Food	mg										µg	
		Na	K	Ca	Mg	P	Fe	Cu	Zn	Cl	Mn	Se	I
96	**Lamb**, average, trimmed lean, raw	70	330	12	22	190	1.4	0.08	3.3	74	0.01	2	6
97	extra trimmed lean, raw	N	N	N	N	N	N	N	N	N	N	N	N
98	trimmed fat, raw	36	140	9	9	86	0.7	0.03	0.9	43	0.01	2	6
99	extra trimmed fat, raw	N	N	N	N	N	N	N	N	N	N	N	N
100	fat, cooked	72	260	11	18	160	1.1	0.05	1.5	67	0.01	4	6
101	**Best end neck cutlets**, raw, lean and fat	58	250	11	17	150	1.0	0.05	1.9	62	0.01	3	8
102	-, -, weighed with bone	42	180	8	12	110	0.7	0.04	1.4	45	0.01	2	6
103	barbecued, lean	82	380	24	26	230	2.0	0.08	3.5	69	0.02	4	6
104	-, -, weighed with fat and bone	42	190	12	13	120	1.0	0.04	1.8	35	0.01	2	3
105	-, lean and fat	80	340	20	24	210	1.8	0.07	2.9	68	0.02	4	6
106	-, -, weighed with bone	57	250	14	17	150	1.3	0.05	2.1	49	0.01	3	4
107	grilled, lean	84	370	23	26	230	1.9	0.11	3.6	71	0.01	4	6
108	-, -, weighed with fat and bone	40	180	11	12	110	0.9	0.05	1.7	34	Tr	2	3
109	-, lean and fat	81	340	19	24	210	1.7	0.09	2.9	71	0.01	4	6
110	-, -, weighed with bone	58	250	14	17	150	1.2	0.06	2.1	51	0.01	3	4
111	**Breast**, raw, lean	86	290	8	20	170	1.2	0.09	3.5	90	0.01	2	8
112	-, lean and fat	69	240	8	16	140	1.0	0.07	2.6	74	0.01	2	8
113	roasted, lean	93	330	8	22	200	1.6	0.07	5.1	67	0.01	4	6
114	-, lean and fat	85	300	9	21	180	1.4	0.06	3.7	67	0.01	4	6

No. 18-	Food	Retinol µg	Carotene µg	Vitamin D µg	Vitamin E mg	Thiamin mg	Ribo-flavin mg	Niacin mg	Trypt 60 mg	Vitamin B6 mg	Vitamin B12 µg	Folate µg	Panto-thenate mg	Biotin µg	Vitamin C mg
96	**Lamb**, average, trimmed lean, *raw*	6	Tr	0.4	0.09	0.09	0.20	5.4	3.9	0.30	2	6	0.92	2	0
97	extra trimmed lean, raw	N	N	N	N	N	N	N	N	N	N	N	N	N	N
98	trimmed fat, *raw*	29	Tr	0.5	0.14	0.07	0.12	2.2	1.3	0.10	1	4	0.47	1	0
99	extra trimmed fat, *raw*	N	N	N	N	N	N	N	N	N	N	N	N	N	N
100	fat, cooked	29	Tr	0.5	0.28	0.09	0.17	3.6	2.0	0.20	1	4	0.74	1	0
101	**Best end neck cutlets**, *raw,* lean and fat	15	Tr	0.4	0.07	0.15	0.16	5.1	3.2	0.34	1	11	0.40	2	0
102	-, -, weighed with bone	11	Tr	0.3	0.05	0.11	0.11	3.7	2.3	0.25	1	8	0.29	1	0
103	*barbecued,* lean	Tr	Tr	0.6	0.12	0.16	0.19	6.6	5.8	0.46	3	11	1.30	2	0
104	-, -, weighed with fat and bone	Tr	Tr	0.3	0.06	0.08	0.10	3.4	2.9	0.23	1	6	0.66	1	0
105	-, lean and fat	8	Tr	0.6	0.16	0.14	0.18	5.8	4.7	0.38	2	9	1.14	2	0
106	-, -, weighed with bone	6	Tr	0.4	0.11	0.10	0.13	4.2	3.4	0.27	1	6	0.82	1	0
107	*grilled,* lean	Tr	Tr	0.6	0.10	0.16	0.19	7.0	5.9	0.40	3	4	1.40	2	0
108	-, -, weighed with fat and bone	Tr	Tr	0.3	0.05	0.08	0.09	3.4	2.9	0.19	1	2	0.67	1	0
109	-, lean and fat	9	Tr	0.6	0.16	0.14	0.19	6.0	4.7	0.32	2	4	1.21	2	0
110	-, -, weighed with bone	7	Tr	0.4	0.11	0.10	0.14	4.3	3.4	0.23	1	3	0.87	1	0
111	**Breast**, *raw,* lean	9	Tr	0.4	0.34	0.09	0.16	5.0	3.1	0.19	2	7	0.90	1	0
112	-, lean and fat	16	Tr	0.4	0.28	0.08	0.14	4.1	2.5	0.16	2	6	0.76	1	0
113	*roasted,* lean	Tr	Tr	0.6	0.11	0.08	0.19	5.7	5.6	0.16	3	6	1.30	2	0
114	-, lean and fat	10	Tr	0.5	0.17	0.09	0.18	4.9	4.2	0.16	2	5	1.09	2	0

Lamb continued

Composition of food per 100g

No. 18-	Food	Description and main data sources	Edible Proportion	Water g	Total Nitrogen g	Protein g	Fat g	Carbo-hydrate g	Energy value kcal	kJ
115	Chump chops, *raw*, lean and fat	Calculated from 77% lean and 23% fat	1.00	62.7	2.92	18.3	18.8	0	242	1007
116	-, -, weighed with bone	Calculated from no. 115	0.87	54.6	2.54	15.9	16.4	0	211	877
117	*fried*, lean	35 samples of a mixture of chops and steaks	1.00	60.0	4.50	28.1	11.2	0	213	892
118	-, -, weighed with fat and bone	Calculated from no. 117	0.63	37.8	2.83	17.7	7.1	0	135	564
119	-, lean and fat	Calculated from 76% lean and 24% fat	1.00	51.1	3.94	24.7	23.2	0	308	1278
120	-, -, weighed with bone	Calculated from no. 119	0.83	42.4	3.27	20.5	19.3	0	256	1063
121	Chump steaks, *raw*, lean and fat	Calculated from 79% lean and 19% fat	1.00	63.8	2.97	18.6	16.4	0	222	923
122	*fried*, lean and fat	Calculated from 80% lean and 19% fat	1.00	51.8	3.97	24.8	20.4	0	283	1176
123	Leg, average, *raw*, lean and fat	Calculated from 83% lean and 17% fat	1.00	67.4	3.05	19.0	12.3	0	187	778
124	Leg chops, *grilled*, lean and fat	Calculated from 91% lean and 9% fat	1.00	59.4	4.53	28.3	12.0	0	221	925
125	-, -, weighed with bone	Calculated from no. 124	0.84	49.9	3.81	23.8	10.1	0	186	778
126	Leg half fillet, *braised*, lean	10 samples	1.00	61.8	4.37	27.3	10.5	0	204	853
127	-, lean and fat	Calculated from 84% lean and 16% fat	1.00	56.9	4.10	25.6	17.1	0	256	1068
128	Leg half knuckle, *pot-roasted*, lean	10 samples	1.00	61.3	4.70	29.4	9.3	0	201	844
129	-, lean and fat	Calculated from 89% lean and 11% fat	1.00	58.1	4.49	28.1	13.8	0	237	988
130	-, -, weighed with bone	Calculated from no. 129	0.73	42.4	3.30	20.5	10.1	0	173	722
131	Leg joint, *roasted*, lean	10 samples; boneless	1.00	58.9	4.93	30.8	9.6	0	210	879
132	-, lean and fat	Calculated from 90% lean and 8% fat	1.00	56.7	4.75	29.7	13.0	0	236	986
133	Leg steaks, *grilled*, lean	29 samples	1.00	61.5	4.67	29.2	9.0	0	198	829
134	-, lean and fat	Calculated from 94% lean and 6% fat	1.00	58.6	4.48	28.0	13.2	0	231	964

No. 18-	Food	Starch g	Total sugars g	Dietary fibre Southgate method g	Englyst method g	Fatty acids *cis & trans* Satd g	Mono-unsatd g	Poly-unsatd g	Total trans g	Cholesterol mg
115	**Chump chops,** *raw,* lean and fat	0	0	0	0	8.8	7.3	1.0	1.6	74
116	-, -, weighed with bone	0	0	0	0	7.6	6.4	0.9	1.4	64
117	*fried,* lean	0	0	0	0	5.0	4.1	0.9	0.7	100
118	-, -, weighed with fat and bone	0	0	0	0	3.2	2.6	0.6	0.4	63
119	-, lean and fat	0	0	0	0	10.8	8.6	1.7	1.7	98
120	-, -, weighed with bone	0	0	0	0	9.0	7.2	1.4	1.4	81
121	**Chump steaks,** *raw,* lean and fat	0	0	0	0	7.6	6.4	0.9	1.4	73
122	*fried,* lean and fat	0	0	0	0	9.4	7.6	1.6	1.4	96
123	**Leg,** average, *raw,* lean and fat	0	0	0	0	5.9	4.8	0.6	1.1	78
124	**Leg chops,** *grilled,* lean and fat	0	0	0	0	5.0	5.1	0.7	0.9	105
125	-, -, weighed with bone	0	0	0	0	4.2	4.3	0.6	0.8	88
126	**Leg half fillet,** *braised,* lean	0	0	0	0	4.6	4.1	0.7	0.6	97
127	-, lean and fat	0	0	0	0	7.7	6.8	1.0	1.2	99
128	**Leg half knuckle,** *pot-roasted,* lean	0	0	0	0	3.9	3.7	0.6	0.6	105
129	-, lean and fat	0	0	0	0	6.0	5.6	0.8	1.0	105
130	-, -, weighed with bone	0	0	0	0	4.4	4.1	0.6	0.8	77
131	**Leg joint,** *roasted,* lean	0	0	0	0	3.4	4.0	0.6	0.7	110
132	-, lean and fat	0	0	0	0	4.7	5.6	0.8	1.0	110
133	**Leg steaks,** *grilled,* lean	0	0	0	0	3.6	3.8	0.6	0.7	105
134	-, lean and fat	0	0	0	0	5.5	5.6	0.8	1.0	105

No. 18-	Food	Na	K	Ca	Mg	P	Fe	Cu	Zn	Cl	Mn	Se	I
							mg					µg	
115	**Chump chops**, *raw*, lean and fat	56	300	15	21	180	1.5	0.07	2.5	65	0.01	2	10
116	-, -, weighed with bone	49	260	13	18	160	1.3	0.06	2.2	56	0.01	2	9
117	*fried*, lean	74	400	33	28	260	2.4	0.12	4.1	70	0.02	4	6
118	-, -, weighed with fat and bone	47	250	21	18	160	1.5	0.08	2.6	44	0.01	3	4
119	-, lean and fat	72	360	27	25	230	2.1	0.10	3.4	68	0.02	4	6
120	-, -, weighed with bone	60	300	22	21	190	1.7	0.08	2.8	56	0.01	3	5
121	**Chump steaks**, *raw*, lean and fat	56	310	15	21	190	1.5	0.07	2.6	65	0.01	2	10
122	*fried*, lean and fat	71	360	28	25	230	2.1	0.10	3.5	67	0.02	4	6
123	**Leg**, average, *raw*, lean and fat	58	320	7	22	190	1.4	0.08	2.8	59	0.01	2	2
124	**Leg chops**, *grilled*, lean and fat	71	410	28	29	250	2.1	0.11	4.5	73	0.02	4	6
125	-, -, weighed with bone	59	340	23	24	210	1.8	0.09	3.8	61	0.02	3	5
126	**Leg half fillet**, *braised*, lean	58	300	9	23	210	2.0	0.10	4.5	68	0.01	4	6
127	-, lean and fat	62	300	10	23	200	1.9	0.09	4.0	69	0.01	4	6
128	**Leg half knuckle**, *pot-roasted*, lean	58	300	11	24	210	2.1	0.11	4.6	74	0.01	4	6
129	-, lean and fat	60	300	11	24	200	2.0	0.10	4.3	74	0.01	4	6
130	-, -, weighed with bone	44	220	8	18	150	1.5	0.07	3.1	54	Tr	3	4
131	**Leg joint**, *roasted*, lean	82	370	10	27	230	2.3	0.12	5.2	77	0.01	4	6
132	-, lean and fat	82	360	10	27	230	2.2	0.12	4.9	77	0.01	4	6
133	**Leg steaks**, *grilled*, lean	70	420	20	29	260	2.2	0.11	4.7	73	0.02	4	6
134	-, lean and fat	71	400	19	28	250	2.1	0.11	4.4	73	0.02	4	6

Lamb *continued*

No. 18-	Food	Retinol μg	Carotene μg	Vitamin D μg	Vitamin E mg	Thiamin mg	Ribo-flavin mg	Niacin mg	Trypt 60 mg	Vitamin B6 mg	Vitamin B12 μg	Folate μg	Panto-thenate mg	Biotin μg	Vitamin C mg
115	**Chump chops**, *raw*, lean and fat	11	Tr	0.5	0.07	0.14	0.20	4.8	3.4	0.19	3	2	1.16	2	0
116	-, -, weighed with bone	10	Tr	0.4	0.06	0.12	0.17	4.2	3.0	0.16	3	2	1.01	1	0
117	*fried*, lean	Tr	Tr	0.6	N	0.18	0.28	6.8	5.9	0.44	3	2	1.30	2	0
118	-, -, weighed with fat and bone	Tr	Tr	0.4	N	0.11	0.18	4.3	3.7	0.28	2	1	0.82	2	0
119	-, lean and fat	7	Tr	0.6	N	0.16	0.25	5.9	4.9	0.37	2	2	1.16	2	0
120	-, -, weighed with bone	6	Tr	0.5	N	0.13	0.21	4.9	4.1	0.31	2	2	0.96	2	0
121	**Chump steaks**, *raw*, lean and fat	10	Tr	0.6	0.07	0.14	0.20	4.9	3.5	0.19	3	2	1.19	2	0
122	*fried*, lean and fat	6	Tr	0.6	N	0.16	0.25	6.0	5.0	0.37	2	2	1.16	2	0
123	**Leg**, average, *raw*, lean and fat	9	Tr	0.7	0.05	0.14	0.23	5.1	3.7	0.33	1	11	1.25	2	0
124	**Leg chops**, *grilled*, lean and fat	Tr	Tr	0.3	0.10	0.20	0.27	7.8	5.8	0.31	3	2	1.36	2	0
125	-, weighed with bone	Tr	Tr	0.2	0.08	0.17	0.23	6.6	4.9	0.26	2	2	1.14	2	0
126	**Leg half fillet**, *braised*, lean	Tr	Tr	0.6	0.01	0.13	0.28	5.7	5.7	0.28	3	8	1.30	2	0
127	-, lean and fat	5	Tr	0.6	0.06	0.13	0.27	5.4	5.1	0.26	2	7	1.22	2	0
128	**Leg half knuckle**, *pot-roasted*, lean	Tr	Tr	0.6	0.01	0.11	0.24	6.0	6.1	0.30	3	3	1.40	3	0
129	-, lean and fat	Tr	Tr	0.6	0.04	0.11	0.23	5.8	5.7	0.30	3	3	1.33	2	0
130	-, -, weighed with bone	Tr	Tr	0.4	0.03	0.08	0.17	4.2	4.2	0.22	2	2	0.97	2	0
131	**Leg joint**, *roasted*, lean	Tr	Tr	0.6	0.06	0.15	0.26	6.2	6.4	0.31	3	6	1.50	2	0
132	-, lean and fat	Tr	Tr	0.6	0.08	0.15	0.25	6.0	6.1	0.30	3	6	1.44	3	0
133	**Leg steaks**, *grilled*, lean	Tr	Tr	0.3	0.08	0.21	0.28	8.1	6.1	0.32	3	2	1.40	3	0
134	-, lean and fat	Tr	Tr	0.3	0.10	0.20	0.27	7.7	5.7	0.31	3	2	1.34	2	0

Lamb continued

Composition of food per 100g

No. 18-	Food	Description and main data sources	Edible Proportion	Water g	Total Nitrogen g	Protein g	Fat g	Carbo-hydrate g	Energy value kcal	kJ
135	**Leg, whole**, *roasted medium*, lean	10 samples	1.00	60.5	4.75	29.7	9.4	0	203	853
136	-, lean and fat	Calculated from 89% lean and 11% fat	1.00	57.3	4.50	28.1	14.2	0	146	602
137	*roasted well-done*, lean	10 samples	1.00	59.1	5.01	31.3	9.2	0	208	873
138	-, lean and fat	Calculated from 89% lean and 11% fat	1.00	56.2	4.77	29.8	13.6	0	242	1010
139	**Loin chops**, *raw*, lean and fat	Calculated from 72% lean and 28% fat	1.00	59.3	2.81	17.6	23.0	0	277	1150
140	-, -, weighed with bone	Calculated from no. 139	0.78	46.3	2.19	13.7	17.9	0	216	895
141	*grilled*, lean	33 samples	1.00	59.6	4.67	29.2	10.7	0	213	892
142	-, -, weighed with fat and bone	Calculated from no. 141	0.61	36.4	2.85	17.8	6.5	0	130	543
143	-, lean and fat	Calculated from 76% lean and 24% fat	1.00	50.5	4.24	26.5	22.1	0	305	1268
144	-, -, weighed with bone	Calculated from no. 143	0.81	40.9	3.44	21.5	17.9	0	247	1028
145	*microwaved*, lean	10 samples	1.00	54.6	5.18	32.4	12.5	0	242	1013
146	-, -, weighed with fat and bone	Calculated from no. 145	0.58	31.7	3.01	18.8	7.3	0	141	590
147	-, lean and fat	Calculated from 72% lean and 28% fat	1.00	45.3	4.39	27.5	26.9	0	352	1463
148	-, -, weighed with bone	Calculated from no. 147	0.82	37.1	3.60	22.5	22.1	0	289	1200
149	*roasted*, lean	35 samples	1.00	52.1	5.50	34.4	13.3	0	257	1077
150	-, -, weighed with fat and bone	Calculated from no. 149	0.56	29.2	3.08	19.3	7.4	0	144	602
151	-, lean and fat	Calculated from 73% lean and 27% fat	1.00	43.8	4.66	29.1	26.9	0	359	1490
152	-, -, weighed with bone	Calculated from no. 151	0.88	38.5	4.10	22.4	20.7	0	316	1312

Lamb *continued*

Composition of food per 100g

No. Food 18-	Starch g	Total sugars g	Dietary fibre Southgate method g	Dietary fibre Englyst method g	Fatty acids Satd g	Fatty acids cis & trans Mono-unsatd g	Fatty acids cis & trans Poly-unsatd g	Fatty acids Total trans g	Cholesterol mg
135 **Leg, whole,** *roasted medium,* lean	0	0	0	0	3.8	3.9	0.6	0.7	100
136 -, lean and fat	0	0	0	0	5.9	6.1	0.8	1.1	100
137 *roasted well-done,* lean	0	0	0	0	3.7	3.9	0.6	0.7	110
138 -, lean and fat	0	0	0	0	5.7	5.8	0.8	1.0	110
139 **Loin chops,** *raw,* lean and fat	0	0	0	0	10.8	8.8	1.2	1.8	79
140 -, -, weighed with bone	0	0	0	0	8.4	6.9	1.0	1.4	62
141 *grilled,* lean	0	0	0	0	4.9	4.0	0.6	0.8	96
142 -, -, weighed with fat and bone	0	0	0	0	3.0	2.5	0.4	0.5	59
143 -, lean and fat	0	0	0	0	10.5	8.4	1.3	1.8	100
144 -, -, weighed with bone	0	0	0	0	8.5	6.7	1.0	1.5	81
145 *microwaved,* lean	0	0	0	0	5.7	4.7	0.7	0.9	115
146 -, -, weighed with fat and bone	0	0	0	0	3.3	2.8	0.4	0.5	67
147 -, lean and fat	0	0	0	0	12.8	10.2	1.5	2.3	110
148 -, -, weighed with bone	0	0	0	0	10.5	8.4	1.2	1.9	90
149 *roasted,* lean	0	0	0	0	6.1	5.0	0.8	1.0	120
150 -, -, weighed with fat and bone	0	0	0	0	3.4	2.8	0.4	0.5	68
151 -, lean and fat	0	0	0	0	12.8	10.2	1.5	2.2	115
152 -, -, weighed with bone	0	0	0	0	11.3	9.0	1.3	1.9	100

Lamb continued

Inorganic constituents per 100g food

No. 18-	Food	Na	K	Ca	Mg	P	Fe	Cu	Zn	Cl	Mn	Se	I
						mg						µg	
135	**Leg, whole,** *roasted medium,* lean	63	360	7	26	220	3.1	0.11	4.6	67	0.02	4	3
136	-, lean and fat	64	340	7	25	210	2.9	0.10	4.3	67	0.02	4	3
137	*roasted well-done,* lean	62	350	8	27	230	2.3	0.13	4.8	78	0.01	4	6
138	-, lean and fat	64	340	9	26	220	2.2	0.12	4.5	78	0.01	4	6
139	**Loin chops,** *raw,* lean and fat	63	280	13	19	170	1.3	0.07	2.0	65	0.01	3	7
140	-, -, weighed with bone	49	220	10	15	130	1.0	0.05	1.6	51	0.01	2	5
141	*grilled,* lean	80	400	22	28	240	2.1	0.10	3.6	73	0.02	4	6
142	-, -, weighed with fat and bone	49	240	13	17	150	1.3	0.06	2.2	45	0.01	2	4
143	-, lean and fat	81	370	20	27	230	1.9	0.09	3.1	74	0.02	4	6
144	-, -, weighed with bone	66	300	16	21	190	1.5	0.07	2.5	60	0.01	3	5
145	*microwaved,* lean	75	340	20	26	220	2.1	0.11	4.0	81	0.01	4	6
146	-, -, weighed with fat and bone	44	190	12	15	130	1.2	0.06	2.3	47	0.01	2	3
147	-, lean and fat	74	310	17	24	200	1.8	0.09	3.3	76	0.01	4	6
148	-, -, weighed with bone	60	250	14	20	160	1.5	0.07	2.7	62	0.01	3	5
149	*roasted,* lean	91	410	23	30	260	2.5	0.13	5.8	86	0.01	4	6
150	-, -, weighed with fat and bone	51	230	13	17	150	1.4	0.07	3.2	48	0.01	2	3
151	-, lean and fat	85	370	20	27	230	2.1	0.11	4.6	80	0.01	4	6
152	-, -, weighed with bone	75	330	18	24	200	1.8	0.10	4.0	70	0.01	3	5

Lamb continued

No. Food 18-	Retinol µg	Carotene µg	Vitamin D µg	Vitamin E mg	Thiamin mg	Ribo-flavin mg	Niacin mg	Trypt 60 mg	Vitamin B6 mg	Vitamin B12 µg	Folate µg	Panto-thenate mg	Biotin µg	Vitamin C mg
135 **Leg, whole,** *roasted medium,* lean	Tr	Tr	0.7	0.03	0.12	0.29	6.2	5.8	0.34	2	2	1.50	3	0
136 -, lean and fat	Tr	Tr	0.6	0.05	0.12	0.28	5.9	5.4	0.32	2	2	1.41	2	0
137 *roasted well-done,* lean	Tr	Tr	0.6	0.04	0.13	0.27	6.9	6.5	0.30	3	3	1.50	3	0
138 -, lean and fat	Tr	Tr	0.6	0.07	0.13	0.26	6.6	6.1	0.29	3	3	1.43	3	0
139 **Loin chops,** *raw,* lean and fat	12	Tr	0.4	0.07	0.13	0.22	5.0	3.6	0.23	1	3	0.86	1	0
140 -, -, weighed with bone	9	Tr	0.3	0.05	0.10	0.17	3.9	2.8	0.18	1	2	0.67	1	0
141 *grilled,* lean	Tr	Tr	0.6	0.02	0.17	0.26	8.3	6.1	0.52	3	6	1.40	3	0
142 -, -, weighed with fat and bone	Tr	Tr	0.4	0.01	0.10	0.16	5.1	3.7	0.32	2	4	0.85	2	0
143 -, lean and fat	7	Tr	0.6	0.09	0.16	0.25	7.3	5.2	0.44	3	6	1.28	2	0
144 -, -, weighed with bone	6	Tr	0.5	0.07	0.13	0.20	5.9	4.2	0.35	2	5	1.04	2	0
145 *microwaved,* lean	Tr	Tr	0.6	0.09	0.15	0.21	6.3	6.7	0.32	3	3	1.60	3	0
146 -, -, weighed with fat and bone	Tr	Tr	0.3	0.05	0.09	0.12	3.7	3.9	0.19	2	2	0.93	2	0
147 -, lean and fat	8	Tr	0.6	0.14	0.14	0.20	5.5	5.4	0.27	3	3	1.34	2	0
148 -, -, weighed with bone	7	Tr	0.5	0.11	0.11	0.16	4.5	4.4	0.22	2	2	1.10	2	0
149 *roasted,* lean	Tr	Tr	0.6	0.06	0.16	0.38	6.9	7.2	0.34	3	6	1.60	3	0
150 -, -, weighed with fat and bone	Tr	Tr	0.3	0.03	0.09	0.21	3.9	4.0	0.19	2	3	0.90	2	0
151 -, lean and fat	8	Tr	0.6	0.11	0.14	0.31	6.0	5.7	0.29	3	5	1.35	2	0
152 -, -, weighed with bone	7	Tr	0.5	0.10	0.12	0.27	5.3	5.0	0.25	2	4	1.18	2	0

Lamb continued

Composition of food per 100g

No. 18-	Food	Description and main data sources	Edible Proportion	Water g	Total Nitrogen g	Protein g	Fat g	Carbo-hydrate g	Energy value kcal	kJ
153	**Loin joint**, *raw*, lean and fat	Calculated from 66% lean and 34% fat	1.00	56.7	2.70	16.9	26.6	0	307	1272
154	-, -, weighed with bone	Calculated from no. 153	0.77	43.7	2.08	13.0	20.5	0	236	979
155	*roasted*, lean	10 samples	1.00	60.4	4.51	28.2	10.7	0	209	875
156	-, lean and fat	Calculated from 78% lean and 22% fat	1.00	51.8	4.04	25.3	22.4	0	303	1259
157	-, -, weighed with bone	Calculated from no. 156	0.74	38.3	2.99	18.7	16.6	0	224	932
158	**Mince**, *raw*	10 samples	1.00	67.1[a]	3.06	19.1	13.3[b]	0	196	817
159	*stewed*	10 samples	1.00	62.8	3.90	24.4	12.3	0	208	870
160	**Neck fillet**, *raw*, lean	10 samples	1.00	69.6	3.10	19.4	13.9	0	203	844
161	-, lean and fat	Calculated from 87% lean and 12% fat	1.00	65.8	2.94	18.4	17.6	0	232	964
162	**slices**, *grilled*, lean	10 samples	1.00	52.3	4.46	27.9	19.1	0	284	1181
163	-, lean and fat	Calculated from 89% lean and 9% fat	1.00	49.1	4.20	26.3	21.9	0	302	1257
164	**strips**, *stir-fried*, lean	10 samples	1.00	55.3	3.90	24.4	20.0	0	278	1155
165	-, lean and fat	Calculated from 84% lean and 14% fat	1.00	51.4	3.72	23.2	23.1	0	301	1249
166	**Rack of lamb**, *raw*, lean and fat	Calculated from 69% lean and 31% fat	1.00	57.3	2.76	17.3	23.8	0	283	1175
167	-, -, weighed with bone	Calculated from no. 166	0.72	41.3	2.00	12.5	17.1	0	204	843
168	*roasted*, lean	10 samples	1.00	57.6	4.34	27.1	13.0	0	225	942
169	-, lean and fat	Calculated from 66% lean and 34% fat	1.00	45.5	3.67	23.0	30.1	0	363	1505
170	**Shoulder**, *raw*, lean and fat	Calculated from 76% lean and 24% fat	1.00	61.6	2.81	17.6	18.3	0	235	976
171	-, -, weighed with bone	Calculated from no. 170	0.80	49.3	2.25	14.1	14.6	0	188	780

[a] Water ranged from 63.0g to 71.6g per 100g

[b] Fat ranged from 8.1g to 22.8g per 100g

Lamb continued

18-153 to 18-171
Composition of food per 100g

No. 18-	Food	Starch g	Total sugars g	Dietary fibre Southgate method g	Englyst method g	Fatty acids Satd g	cis & trans Mono-unsatd g	Poly-unsatd g	Total trans g	Cholesterol mg
153	**Loin joint**, *raw*, lean and fat	0	0	0	0	12.6	10.2	1.4	2.1	80
154	-, -, weighed with bone	0	0	0	0	9.7	7.8	1.1	1.6	62
155	*roasted*, lean	0	0	0	0	4.9	4.0	0.6	0.8	100
156	-, lean and fat	0	0	0	0	10.6	8.5	1.3	1.8	100
157	-, -, weighed with bone	0	0	0	0	7.8	6.3	1.0	1.3	74
158	**Mince**, *raw*	0	0	0	0	6.2	5.3	0.6	1.0	77
159	*stewed*	0	0	0	0	5.8	4.8	0.6	0.8	96
160	**Neck fillet**, *raw*, lean	0	0	0	0	6.4	5.3	0.7	1.1	75
161	-, lean and fat	0	0	0	0	8.3	6.8	0.9	1.5	79
162	**slices**, *grilled*, lean	0	0	0	0	8.7	7.3	0.9	1.5	99
163	-, lean and fat	0	0	0	0	10.3	8.3	1.0	1.8	97
164	**strips**, *stir-fried*, lean	0	0	0	0	8.2	7.6	2.2	1.2	86
165	-, lean and fat	0	0	0	0	9.9	8.9	2.3	1.6	91
166	**Rack of lamb**, *raw*, lean and fat	0	0	0	0	11.5	8.9	1.2	1.9	75
167	-, -, weighed with bone	0	0	0	0	8.3	6.4	0.9	1.4	54
168	*roasted*, lean	0	0	0	0	6.2	4.9	0.6	0.9	96
169	-, lean and fat	0	0	0	0	14.8	11.3	1.4	2.2	97
170	**Shoulder**, *raw*, lean and fat	0	0	0	0	8.5	7.1	1.0	1.0	76
171	-, -, weighed with bone	0	0	0	0	6.8	5.7	0.8	0.8	61

Lamb *continued*

Inorganic constituents per 100g food

No. 18-	Food	Na	K	Ca	Mg	P	Fe	Cu	Zn	Cl	Mn	Se	I
						mg						µg	
153	**Loin joint**, *raw*, lean and fat	61	270	13	18	160	1.3	0.06	2.0	63	0.01	3	7
154	-, -, weighed with bone	47	200	10	14	120	1.0	0.05	1.5	48	0.01	2	5
155	*roasted*, lean	76	350	15	25	220	2.0	0.12	3.7	71	0.02	4	6
156	-, lean and fat	75	330	14	23	200	1.8	0.10	3.2	70	0.02	4	6
157	-, weighed with bone	76	240	10	17	150	1.3	0.08	2.4	52	0.01	3	4
158	**Mince**, *raw*	69	310	17	21	190	1.6	0.08	3.5	68	0.01	2	6
159	*stewed*	59	270	15	20	180	2.1	0.11	4.6	46	0.02	3	5
160	**Neck fillet**, *raw*, lean	61	330	4	20	190	1.2	0.07	4.2	69	0.01	2	6
161	-, lean and fat	59	310	5	19	170	1.2	0.07	3.8	67	0.01	2	6
162	**slices**, *grilled*, lean	78	390	10	25	230	1.9	0.09	6.4	70	0.02	4	6
163	-, lean and fat	76	370	10	24	220	1.8	0.08	5.8	69	0.02	4	6
164	**strips**, *stir-fried*, lean	68	360	7	23	210	1.8	0.08	5.2	61	0.02	4	6
165	-, lean and fat	70	350	8	23	200	1.7	0.08	4.6	63	0.02	4	6
166	**Rack of lamb**, *raw*, lean and fat	61	260	12	18	160	1.0	0.06	2.0	65	0.01	3	8
167	-, -, weighed with bone	44	190	9	13	120	0.7	0.04	1.4	47	0.01	2	6
168	*roasted*, lean	77	330	13	24	220	1.8	0.10	3.6	68	0.01	4	6
169	-, lean and fat	75	310	12	22	190	1.6	0.08	2.8	67	0.01	4	6
170	**Shoulder**, *raw*, lean and fat	63	280	6	19	160	1.1	0.08	3.4	63	0.02	3	3
171	-, -, weighed with bone	50	230	5	15	130	0.9	0.06	2.7	50	0.01	2	2

Lamb *continued*

No. 18-	Food	Retinol µg	Carotene µg	Vitamin D µg	Vitamin E mg	Thiamin mg	Ribo-flavin mg	Niacin mg	Trypt 60 mg	Vitamin B6 mg	Vitamin B12 µg	Folate µg	Panto-thenate mg	Biotin µg	Vitamin C mg
153	**Loin joint,** *raw,* lean and fat	13	Tr	0.4	0.07	0.12	0.22	4.8	3.4	0.21	1	3	0.83	1	0
154	-, -, weighed with bone	10	Tr	0.3	0.06	0.09	0.17	3.7	2.6	0.16	1	2	0.64	1	0
155	*roasted,* lean	Tr	Tr	0.6	0.01	0.14	0.25	7.0	5.9	0.36	3	5	1.40	2	0
156	-, lean and fat	6	Tr	0.6	0.06	0.13	0.23	6.2	5.0	0.31	2	5	1.24	2	0
157	-, -, weighed with bone	Tr	Tr	0.4	0.04	0.10	0.17	3.7	3.6	0.23	2	4	0.92	2	0
158	**Mince,** *raw*	5	Tr	0.6	0.18	0.12	0.18	4.8	3.7	0.20	2	2	0.90	2	0
159	*stewed*	5	Tr	0.5	0.11	0.09	0.21	5.2	5.3	0.21	2	9	0.90	4	0
160	**Neck fillet,** *raw,* lean	5	Tr	0.4	0.03	0.15	0.16	4.1	3.8	0.22	2	1	0.90	2	0
161	-, lean and fat	8	Tr	0.4	0.05	0.14	0.16	3.9	3.5	0.21	2	1	0.86	2	0
162	**slices,** *grilled,* lean	Tr	Tr	0.6	0.12	0.16	0.24	5.1	5.8	0.09	3	7	1.30	2	0
163	-, lean and fat	Tr	Tr	0.6	0.13	0.15	0.23	4.9	5.4	0.09	2	7	1.23	2	0
164	**strips,** *stir-fried,* lean	Tr	Tr	0.6	0.59	0.17	0.20	4.6	5.1	0.20	2	7	1.20	2	0
165	-, lean and fat	Tr	Tr	0.6	0.54	0.16	0.20	4.5	4.6	0.19	2	6	1.13	2	0
166	**Rack of lamb,** *raw,* lean and fat	13	Tr	0.3	0.06	0.16	0.16	5.4	3.4	0.37	2	12	0.40	2	0
167	-, -, weighed with bone	9	Tr	0.2	0.04	0.11	0.11	3.9	2.5	0.27	1	9	0.29	1	0
168	*roasted,* lean	Tr	Tr	0.6	0.07	0.13	0.20	7.2	5.6	0.39	3	6	1.30	2	0
169	-, lean and fat	10	Tr	0.6	0.13	0.12	0.19	5.9	4.4	0.31	2	5	1.09	2	0
170	**Shoulder,** *raw,* lean and fat	11	Tr	0.4	0.13	0.14	0.16	4.4	2.9	0.23	2	4	0.88	1	0
171	-, -, weighed with bone	9	Tr	0.3	0.10	0.11	0.13	3.5	2.3	0.18	1	3	0.70	1	0

Lamb continued

No. 18-	Food	Description and main data sources	Edible Proportion	Water g	Total Nitrogen g	Protein g	Fat g	Carbo-hydrate g	Energy value kcal	Energy value kJ
172	**Shoulder, diced, kebabs,** grilled, lean and fat	Calculated from 85% lean and 15% fat	1.00	52.1	4.56	28.5	19.3	0	288	1199
173	**half bladeside**, pot-roasted, lean	10 samples	1.00	58.3	4.19	26.2	14.3	0	234	975
174	-, lean and fat	Calculated from 72% lean and 28% fat	1.00	50.1	3.76	23.5	25.6	0	324	1347
175	**half knuckle**, braised, lean	10 samples	1.00	62.5	4.02	25.1	12.4	0	212	886
176	-, lean and fat	Calculated from 74% lean and 26% fat	1.00	53.7	3.66	22.9	23.4	0	302	1255
177	roasted, lean	10 samples	1.00	56.7	4.54	28.4	13.5	0	235	982
178	-, lean and fat	Calculated from 82% lean and 16% fat	1.00	51.1	4.15	25.9	19.8	0	282	1173
179	**whole**, roasted, lean	10 samples	1.00	56.9	4.35	27.2	12.1	0	218	910
180	-, lean and fat	Calculated from 78% lean and 22% fat	1.00	50.5	3.96	24.7	22.1	0	298	1238
181	-, -, weighed with bone	Calculated from no. 180	0.79	38.9	3.05	19.5	17.5	0	235	979
182	**Stewing lamb**, raw, lean and fat	Calculated from 80% lean and 20% fat	1.00	63.6	3.60	22.5	12.6	0	203	849
183	-, -, weighed with bone	Calculated from no. 182	0.67	42.6	2.42	15.1	8.4	0	136	568
184	pressure cooked, lean	10 samples	1.00	56.6	4.59	28.7	14.8	0	248	1036
185	-, lean and fat	Calculated from 82% lean and 18% fat	1.00	53.5	4.11	25.7	20.9	0	291	1210
186	stewed, lean	10 samples	1.00	58.9	4.26	26.6	14.8	0	240	1000
187	-, lean and fat	Calculated from 85% lean and 15% fat	1.00	56.1	3.91	24.4	20.1	0	279	1159

Lamb continued

No. 18-	Food	Starch g	Total sugars g	Dietary fibre Southgate method g	Englyst method g	Fatty acids Satd g	cis & trans Mono-unsatd g	Poly-unsatd g	Total trans g	Cholesterol mg
172	**Shoulder, diced, kebabs,** *grilled*, lean and fat	0	0	0	0	9.0	7.5	1.0	1.5	110
173	**half bladeside**, *pot-roasted*, lean	0	0	0	0	6.4	5.5	0.8	1.1	93
174	-, lean and fat	0	0	0	0	12.0	10.0	1.4	2.0	98
175	**half knuckle**, *braised*, lean	0	0	0	0	5.4	4.9	0.7	0.9	89
176	-, lean and fat	0	0	0	0	10.8	9.3	1.3	1.8	94
177	*roasted*, lean	0	0	0	0	6.2	5.2	0.6	0.9	100
178	-, lean and fat	0	0	0	0	9.3	7.8	0.9	1.5	100
179	**whole**, *roasted*, lean	0	0	0	0	5.5	4.7	0.6	0.8	105
180	-, lean and fat	0	0	0	0	10.4	8.7	1.0	1.7	105
181	-, -, weighed with bone	0	0	0	0	8.2	6.9	0.8	1.3	83
182	**Stewing lamb**, *raw*, lean and fat	0	0	0	0	5.9	4.9	0.6	1.0	78
183	-, -, weighed with bone	0	0	0	0	3.9	3.3	0.4	0.7	52
184	*pressure cooked*, lean	0	0	0	0	6.5	5.6	1.0	1.0	100
185	-, lean and fat	0	0	0	0	9.6	8.0	1.4	1.5	98
186	*stewed*, lean	0	0	0	0	6.5	5.6	1.0	1.0	94
187	-, lean and fat	0	0	0	0	9.2	7.7	1.3	1.4	92

Lamb continued

Inorganic constituents per 100g food

No. 18-	Food	Na	K	Ca	Mg	P	Fe	Cu	Zn	Cl	Mn	Se	I
						mg						µg	
172	**Shoulder, diced, kebabs,** grilled, lean and fat	87	420	14	28	240	1.8	0.12	5.6	76	0.03	4	6
173	**half bladeside**, pot-roasted, lean	68	280	12	21	190	1.8	0.12	5.9	66	0.01	4	6
174	-, lean and fat	71	280	12	21	180	1.6	0.10	4.7	68	0.01	4	6
175	**half knuckle**, braised, lean	76	290	10	21	190	1.6	0.09	5.1	63	0.01	4	6
176	-, lean and fat	76	290	11	21	180	1.5	0.08	4.2	65	0.01	4	6
177	roasted, lean	75	340	9	25	210	2.1	0.11	4.8	71	0.01	4	6
178	-, lean and fat	74	330	9	24	200	1.9	0.10	4.2	70	0.01	4	6
179	**whole**, roasted, lean	80	330	8	23	210	1.8	0.10	5.8	81	0.01	6	6
180	-, lean and fat	80	320	9	22	190	1.6	0.09	5.0	79	0.01	5	7
181	-, -, weighed with bone	63	260	7	17	150	1.3	0.07	3.9	62	0.01	4	5
182	**Stewing lamb**, raw, lean and fat	66	270	25	17	160	1.2	0.06	3.1	69	0.01	2	4
183	-, -, weighed with bone	44	180	17	11	110	0.8	0.04	2.1	46	0.01	1	3
184	pressure cooked, lean	65	240	31	20	180	1.9	0.13	6.0	72	0.01	4	6
185	-, lean and fat	63	240	27	19	170	1.7	0.11	5.1	68	0.01	4	6
186	stewed, lean	49	160	37	16	140	1.9	0.09	6.1	67	0.01	4	6
187	-, lean and fat	50	170	33	16	140	1.7	0.08	5.4	65	0.01	4	6

Lamb *continued*

No. 18-	Food	Retinol μg	Carotene μg	Vitamin D μg	Vitamin E mg	Thiamin mg	Ribo-flavin mg	Niacin mg	Trypt 60 mg	Vitamin B6 mg	Vitamin B12 μg	Folate μg	Panto-thenate mg	Biotin μg	Vitamin C mg
172	**Shoulder, diced, kebabs,** *grilled*, lean and fat	Tr	Tr	0.6	0.19	0.12	0.25	6.8	5.7	0.21	3	7	1.40	2	0
173	**half bladeside**, *pot-roasted*, lean	Tr	Tr	0.6	0.19	0.07	0.19	4.2	5.5	0.12	2	5	1.30	2	0
174	-, lean and fat	8	Tr	0.6	0.22	0.08	0.19	4.1	4.5	0.13	2	5	1.16	2	0
175	**half knuckle**, *braised*, lean	Tr	Tr	0.6	0.09	0.11	0.23	5.1	5.2	0.24	2	2	1.20	2	0
176	-, lean and fat	8	Tr	0.6	0.14	0.11	0.22	4.8	4.4	0.22	2	3	1.10	2	0
177	*roasted*, lean	Tr	Tr	0.6	0.05	0.13	0.24	5.7	5.9	0.28	3	5	1.40	2	0
178	-, lean and fat	5	Tr	0.6	0.09	0.12	0.23	5.3	5.2	0.26	2	5	1.28	2	0
179	**whole**, *roasted*, lean	Tr	Tr	0.8	0.06	0.11	0.23	5.3	5.6	0.21	2	4	1.10	2	0
180	-, lean and fat	6	Tr	0.7	0.15	0.10	0.21	5.0	4.8	0.20	2	4	0.99	2	0
181	-, -, weighed with bone	5	Tr	0.5	0.12	0.08	0.17	3.9	3.8	0.16	2	3	0.78	1	0
182	**Stewing lamb**, *raw*, lean and fat	10	Tr	0.4	0.07	0.09	0.15	3.8	3.2	0.23	2	2	0.40	1	0
183	-, -, weighed with bone	7	Tr	0.3	0.05	0.06	0.10	2.5	2.1	0.15	1	1	0.27	1	0
184	*pressure cooked*, lean	Tr	Tr	0.6	0.04	0.14	0.20	4.3	6.0	0.18	3	2	1.40	2	0
185	-, lean and fat	5	Tr	0.6	0.07	0.13	0.19	4.0	5.2	0.17	3	2	1.26	2	0
186	*stewed*, lean	Tr	Tr	0.6	0.20	0.04	0.12	2.3	5.5	0.11	3	2	1.30	2	0
187	-, lean and fat	Tr	Tr	0.6	0.20	0.05	0.12	2.4	4.9	0.11	2	2	1.19	2	0

Lamb continued

Composition of food per 100g

New Zealand lamb

No. 18-	Food	Description and main data sources	Edible Proportion	Water g	Total Nitrogen g	Protein g	Fat g	Carbohydrate g	Energy value kcal	kJ
188	Leg, whole, chilled, *roasted*, lean	7 samples	1.00	60.4	4.61	28.8	10.2	0	207	867
189	-, lean and fat	Calculated from 92% lean and 8% fat	1.00	58.3	4.48	28.0	13.1	0	230	961
190	frozen, raw, lean and fat, weighed with bone	Calculated from 64% lean and 12% fat	0.77	50.8	2.40	15.0	11.1	0	160	666
191	-, *roasted*, lean	10 samples	1.00	60.0	4.90	30.6	9.3	0	206	864
192	-, lean and fat	Calculated from 89% lean and 11% fat	1.00	57.2	4.64	29.0	14.1	0	243	1015
193	-, -, weighed with bone	Calculated from no. 192	0.75	41.7	3.38	21.2	10.3	0	178	742
194	Loin chops, frozen, *grilled*, lean	44 samples	1.00	54.6	5.02	31.4	13.7	0	249	1041
195	-, -, weighed with fat and bone	Calculated from no. 194	0.59	32.2	2.96	18.5	8.1	0	147	614
196	-, lean and fat	Calculated from 75% lean and 25% fat	1.00	44.4	4.30	26.9	28.5	0	364	1512
197	-, -, weighed with bone	Calculated from no. 196	0.79	35.1	3.40	21.2	22.5	0	287	1193
198	Mince, frozen, *stewed*	8 samples	1.00	61.8	3.34	20.9	17.7	0	243	1010
199	Shoulder, whole, frozen, *roasted*, lean	9 samples	1.00	56.3	4.42	27.6	15.6	0	251	1046
200	-, lean and fat	Calculated from 75% lean and 25% fat	1.00	47.9	3.93	24.6	27.9	0	350	1451

Lamb *continued*

Composition of food per 100g

No. 18-	Food	Starch g	Total sugars g	Dietary fibre Southgate method g	Dietary fibre Englyst method g	Fatty acids *cis & trans* Satd g	Mono-unsatd g	Poly-unsatd g	Total trans g	Cholesterol mg
	New Zealand lamb									
188	**Leg, whole**, chilled, *roasted*, lean	0	0	0	0	4.4	4.0	0.7	0.7	110
189	-, lean and fat	0	0	0	0	5.8	5.2	0.9	0.9	110
190	frozen, *raw*, lean and fat, weighed with bone	0	0	0	0	3.7	3.4	0.6	0.6	58
191	-, *roasted*, lean	0	0	0	0	4.0	3.7	0.6	0.6	110
192	-, lean and fat	0	0	0	0	6.5	5.9	1.0	1.0	110
193	-, -, weighed with bone	0	0	0	0	4.9	4.4	0.7	0.7	82
194	**Loin chops**, frozen, *grilled*, lean	0	0	0	0	6.2	5.3	0.7	1.0	110
195	-, -, weighed with fat and bone	0	0	0	0	3.7	3.1	0.4	0.6	65
196	-, lean and fat	0	0	0	0	13.5	11.0	1.5	2.3	105
197	-, -, weighed with bone	0	0	0	0	10.7	8.7	1.2	1.8	83
198	**Mince**, frozen, *stewed*	0	0	0	0	8.8	6.5	0.9	1.3	110
199	**Shoulder, whole**, frozen, *roasted*, lean	0	0	0	0	7.1	5.9	0.9	1.1	97
200	-, lean and fat	0	0	0	0	13.3	10.7	1.5	2.1	98

Lamb *continued*

Inorganic constituents per 100g food

New Zealand lamb

No. 18-	Food	mg										µg	
		Na	K	Ca	Mg	P	Fe	Cu	Zn	Cl	Mn	Se	I
188	**Leg, whole,** chilled, *roasted,* lean	82	350	15	26	220	2.1	0.13	4.5	72	0.01	4	6
189	-, lean and fat	82	350	15	26	210	2.0	0.12	4.3	72	0.01	4	6
190	frozen, *raw,* lean and fat, weighed with bone	44	240	3	17	140	1.2	0.07	2.3	44	0.01	1	3
191	-, *roasted,* lean	62	340	6	27	230	2.5	0.13	5.3	77	0.02	4	6
192	-, lean and fat	63	340	7	26	220	2.4	0.12	4.8	76	0.02	4	6
193	-, -, weighed with bone	47	250	5	19	170	1.8	0.09	3.6	57	0.01	3	4
194	**Loin chops,** frozen, *grilled,* lean	80	400	19	30	260	2.3	0.13	4.3	79	0.01	4	6
195	-, -, weighed with fat and bone	60	240	11	18	150	1.4	0.08	2.5	47	0.01	2	4
196	-, lean and fat	76	360	17	27	230	2.0	0.11	3.5	74	0.01	4	6
197	-, -, weighed with bone	60	290	13	21	180	1.6	0.09	2.8	58	0.01	3	5
198	**Mince,** frozen, *stewed*	75	230	12	18	170	3.2	0.18	3.5	52	0.02	6	4
199	**Shoulder, whole,** frozen, *roasted,* lean	81	340	5	25	210	1.9	0.11	6.0	80	0.01	4	4
200	-, lean and fat	79	320	6	24	200	1.7	0.10	4.9	77	0.01	4	4

No. 18-	Food	Retinol µg	Carotene µg	Vitamin D µg	Vitamin E mg	Thiamin mg	Ribo- flavin mg	Niacin mg	Trypt 60 mg	Vitamin B6 mg	Vitamin B12 µg	Folate µg	Panto- thenate mg	Biotin µg	Vitamin C mg
	New Zealand lamb														
188	**Leg, whole,** chilled, *roasted*, lean	Tr	Tr	0.6	0.10	0.15	0.32	5.1	6.0	0.38	3	10	1.40	3	0
189	-, lean and fat	Tr	Tr	0.6	0.11	0.15	0.31	5.0	5.7	0.36	3	10	1.36	2	0
190	frozen, *raw*, lean and fat, weighed with bone	Tr	Tr	0.4	0.03	0.18	0.16	3.1	3.0	0.29	2	6	0.82	1	0
191	-, *roasted*, lean	Tr	Tr	0.6	0.07	0.14	0.27	4.6	6.4	0.26	3	5	1.50	3	0
192	-, lean and fat	Tr	Tr	0.6	0.10	0.10	0.26	4.6	5.9	0.25	3	5	1.42	2	0
193	-, -, weighed with bone	Tr	Tr	0.4	0.07	0.07	0.19	3.4	4.4	0.19	2	4	1.04	1	0
194	**Loin chops,** frozen, *grilled*, lean	Tr	Tr	0.6	0.11	0.17	0.33	7.1	6.5	0.36	3	6	1.50	3	0
195	-, -, weighed with fat and bone	Tr	Tr	0.3	0.06	0.10	0.19	4.2	3.8	0.21	2	4	0.89	2	0
196	-, lean and fat	7	Tr	0.6	0.14	0.15	0.28	6.1	5.3	0.30	3	5	1.27	2	0
197	-, -, weighed with bone	6	Tr	0.5	0.11	0.12	0.22	4.8	4.2	0.24	2	4	1.00	2	0
198	**Mince,** frozen, *stewed*	5	Tr	0.6	0.29	0.16	0.34	4.2	4.4	0.12	2	6	1.00	2	0
199	**Shoulder, whole,** frozen, *roasted*, lean	Tr	Tr	0.7	0.07	0.11	0.25	3.8	5.8	0.17	3	5	1.70	2	0
200	-, lean and fat	7	Tr	0.6	0.12	0.11	0.23	3.8	4.9	0.17	3	5	1.45	2	0

53

Pork

Composition of food per 100g

No. 18-	Food	Description and main data sources	Edible Proportion	Water g	Total Nitrogen g	Protein g	Fat g	Carbohydrate g	Energy value kcal	Energy value kJ
201	**Pork**, average, trimmed lean, *raw*	LGC; average of 8 different cuts	1.00	74.0	3.49	21.8	4.0	0	123	519
202	extra trimmed lean, *raw*	MLC; weighted average of 6 different cuts	1.00	72.2	3.50	21.9	3.7	0	121	509
203	trimmed fat, *raw*	LGC; average of 8 different cuts	1.00	33.6	1.62	10.1	56.4	0	548	2259
204	extra trimmed fat, *raw*	MLC; weighted average of 6 different cuts	1.00	23.2	1.13	7.1	69.7	0	656	2700
205	fat, *cooked*	LGC; average of 5 different cuts	1.00	33.1	2.27	14.2	50.9	0	515	2125
206	**Belly joint/slices**, *raw*, lean and fat	12 samples	1.00	61.7	3.06	19.1	20.2	0	258	1072
207	-, -, weighed with bone	Calculated from no. 206	0.92	56.8	2.81	17.6	18.6	0	238	987
208	*roasted*, lean and fat	10 samples, 65% lean and 35% fat	1.00	55.8	4.02	25.1	21.4	0	293	1219
209	*grilled*, lean and fat	25 samples, 58% lean and 42% fat	1.00	48.6	4.38	27.4	23.4	0	320	1332
210	**Chump chops**, *raw*, lean and fat	Calculated from 80% lean and 20% fat	1.00	65.0	3.22	20.1	12.6	0	194	808
211	-, -, weighed with bone	Calculated from no. 210	0.86	55.9	2.77	17.3	9.8	0	194	694
212	*fried*, lean and fat	Calculated from 74% lean and 26% fat	1.00	53.6	3.93	24.6	21.6	0	293	1217
213	-, -, weighed with bone	Calculated from no. 212	0.85	45.5	3.34	20.9	18.4	0	249	1036
214	**Chump steaks**, *raw*, lean and fat	Calculated from 91% lean and 9% fat	1.00	69.1	3.41	21.3	7.3	0	151	632
215	*fried*, lean and fat	Calculated from 91% lean and 9% fat	1.00	60.5	4.46	27.9	11.7	0	217	907
216	**Crackling**, *cooked*	From various cuts of pork, grilled and roasted	1.00	16.9	5.79	36.2	45.0	0	550	2280
217	**Diced**, *raw*, lean	10 samples	1.00	74.5	3.42	21.4	4.0	0	122	512
218	-, lean and fat	Calculated from 91% lean and 8% fat	1.00	71.2	3.28	20.5	7.2	0	147	615
219	*casseroled*, lean	10 samples	1.00	62.2	5.07	31.7	6.4	0	184	776
220	-, lean and fat	Calculated from 96% lean and 2% fat	1.00	60.7	4.92	30.8	6.8	0	184	775

No. 18-	Food	Starch g	Total sugars g	Dietary fibre Southgate method g	Dietary fibre Englyst method g	Fatty acids cis & trans Satd g	Fatty acids cis & trans Mono-unsatd g	Fatty acids cis & trans Poly-unsatd g	Total trans g	Cholesterol mg
201	**Pork**, average, trimmed lean, *raw*	0	0	0	0	1.4	1.5	0.7	Tr	63
202	extra trimmed lean, *raw*	0	0	0	0	N	N	N	N	N
203	trimmed fat, *raw*	0	0	0	0	20.4	23.7	9.5	0.3	71
204	extra trimmed fat, *raw*	0	0	0	0	N	N	N	N	N
205	fat, *cooked*	0	0	0	0	17.9	21.5	8.9	0.2	98
206	**Belly joint/slices**, *raw*, lean and fat	0	0	0	0	7.3	8.4	3.1	0.1	71
207	-, -, weighed with bone	0	0	0	0	6.7	7.7	2.8	0.1	65
208	*roasted*, lean and fat	0	0	0	0	7.4	8.5	3.9	0.1	85
209	*grilled*, lean and fat	0	0	0	0	8.2	9.5	4.0	0.1	97
210	**Chump chops**, *raw*, lean and fat	0	0	0	0	4.5	5.0	2.1	0.1	66
211	-, -, weighed with bone	0	0	0	0	3.9	4.3	1.8	0.1	57
212	*fried*, lean and fat	0	0	0	0	7.0	8.3	4.6	0.1	84
213	-, -, weighed with bone	0	0	0	0	5.9	7.0	3.9	0.1	71
214	**Chump steaks**, *raw*, lean and fat	0	0	0	0	2.6	2.9	1.2	Tr	66
215	*fried*, lean and fat	0	0	0	0	3.7	4.4	2.6	0.1	86
216	**Crackling**, *cooked*	0	0	0	0	15.3	19.6	7.9	0.2	105
217	**Diced**, *raw*, lean	0	0	0	0	1.4	1.6	0.7	Tr	63
218	-, lean and fat	0	0	0	0	2.5	2.9	1.2	Tr	65
219	*casseroled*, lean	0	0	0	0	1.9	2.3	1.6	Tr	99
220	-, lean and fat	0	0	0	0	2.0	2.4	1.7	Tr	92

Pork

Inorganic constituents per 100g food

No. 18-	Food	Na	K	Ca	Mg	P	Fe	Cu	Zn	Cl	Mn	Se	I
						mg						µg	
201	Pork, average, trimmed lean, *raw*	63	380	7	24	190	0.7	0.05	2.1	51	0.01	13	5
202	extra trimmed lean, *raw*	N	N	N	N	N	N	N	N	N	N	N	N
203	trimmed fat, *raw*	47	160	9	9	91	0.4	0.04	0.6	51	Tr	7	5
204	extra trimmed fat, *raw*	N	N	N	N	N	N	N	N	N	N	N	N
205	fat, *cooked*	69	240	10	14	140	0.6	0.05	0.9	67	Tr	9	5
206	Belly joint/slices, *raw*, lean and fat	69	290	6	17	160	0.6	0.08	1.7	67	0.02	12	5
207	-, -, weighed with bone	75	270	6	16	150	0.6	0.07	1.6	62	0.02	11	5
208	*roasted*, lean and fat	90	310	9	20	190	0.7	0.06	2.5	93	Tr	15	12
209	*grilled*, lean and fat	97	350	20	23	220	0.9	0.12	2.9	96	0.02	17	5
210	Chump chops, *raw*, lean and fat	54	360	8	22	200	0.6	0.02	1.5	50	Tr	12	3
211	-, -, weighed with bone	43	300	7	19	170	0.5	0.02	1.3	39	Tr	10	3
212	*fried*, lean and fat	60	370	13	24	220	0.8	0.04	1.9	58	Tr	15	3
213	-, -, weighed with bone	51	310	11	21	190	0.7	0.04	1.6	49	Tr	13	3
214	Chump steaks, *raw*, lean and fat	56	380	8	24	210	0.7	0.02	1.6	51	Tr	12	3
215	*fried*, lean and fat	62	410	14	27	250	0.9	0.04	2.2	59	Tr	17	3
216	Crackling, *cooked*	N	N	N	N	N	N	N	N	N	N	N	N
217	Diced, *raw*, lean	70	380	7	24	220	0.8	0.08	2.2	50	0.02	13	5
218	-, lean and fat	68	360	7	23	210	0.8	0.08	2.1	50	0.02	12	5
219	*casseroled*, lean	37	220	12	21	180	1.2	0.13	3.6	39	0.02	20	5
220	-, lean and fat	37	220	12	20	180	1.2	0.13	3.5	39	0.02	19	5

Pork

Vitamins per 100g food

No. 18-	Food	Retinol µg	Carotene µg	Vitamin D µg	Vitamin E mg	Thiamin mg	Ribo-flavin mg	Niacin mg	Trypt 60 mg	Vitamin B6 mg	Vitamin B12 µg	Folate µg	Panto-thenate mg	Biotin µg	Vitamin C mg
201	Pork, average, trimmed lean, raw	Tr	Tr	0.5	0.05	0.98	0.24	6.9	4.5	0.54	1	3	1.46	2	0
202	extra trimmed lean, raw	N	N	N	N	N	N	N	N	N	N	N	N	N	0
203	trimmed fat, raw	Tr	N	1.3	0.03	0.20	0.13	2.1	1.1	0.11	1	2	0.61	5	0
204	extra trimmed fat, raw	N	N	N	N	N	N	N	N	N	N	N	N	N	0
205	fat, cooked	Tr	Tr	2.1	0.05	0.37	0.16	3.8	1.5	0.16	Tr	2	0.86	8	0
206	Belly joint/slices, raw, lean and fat	Tr	Tr	0.8	0.12	0.43	0.22	5.0	3.4	0.39	1	8	1.31	1	0
207	-, -, weighed with bone	Tr	Tr	0.7	0.11	0.40	0.20	4.6	3.1	0.36	1	7	1.21	1	0
208	roasted, lean and fat	Tr	Tr	1.0	0.01	0.47	0.19	8.4	4.5	0.30	1	3	1.42	4	0
209	grilled, lean and fat	Tr	Tr	1.1	0.03	0.60	0.18	7.0	4.9	0.38	1	8	1.77	4	0
210	Chump chops, raw, lean and fat	Tr	Tr	0.7	0.06	0.78	0.18	6.9	4.6	0.34	1	6	1.46	3	0
211	-, -, weighed with bone	Tr	Tr	0.6	0.05	0.67	0.15	5.9	4.0	0.29	1	5	1.26	2	0
212	fried, lean and fat	Tr	Tr	0.9	N	0.76	0.21	7.6	4.8	0.38	Tr	2	2.68	5	0
213	-, -, weighed with bone	Tr	Tr	0.8	N	0.64	0.17	6.5	4.1	0.32	Tr	2	2.28	4	0
214	Chump steaks, raw, lean and fat	Tr	Tr	0.6	0.07	0.85	0.18	7.4	5.0	0.36	1	7	1.56	2	0
215	fried, lean and fat	Tr	Tr	0.8	N	0.89	0.22	8.8	5.7	0.44	Tr	2	3.17	5	0
216	Crackling, cooked	Tr	Tr	N	N	N	N	N	N	N	N	N	N	N	0
217	Diced, raw, lean	Tr	Tr	0.5	0.02	1.04	0.25	7.0	4.4	0.51	1	1	1.47	2	0
218	-, lean and fat	Tr	Tr	0.6	0.02	0.97	0.24	6.6	4.2	0.48	1	1	1.40	2	0
219	casseroled, lean	Tr	Tr	0.8	0.05	0.48	0.25	4.2	6.6	0.36	1	3	0.94	5	0
220	-, lean and fat	Tr	Tr	0.8	0.05	0.47	0.24	4.1	6.3	0.35	1	3	0.92	5	0

Pork continued

No. 18-	Food	Description and main data sources	Edible Proportion	Water g	Total Nitrogen g	Protein g	Fat g	Carbo-hydrate g	Energy value kcal	kJ
221	Diced, kebabs, grilled, lean	10 samples	1.00	60.8	5.46	34.1	4.7	0	179	754
222	-, lean and fat	Calculated from 96% lean and 4% fat	1.00	59.9	5.37	33.6	6.1	0	189	797
223	slow-cooked, lean	10 samples	1.00	64.1	4.80	30.0	5.4	0	169	710
224	-, lean and fat	Calculated from 98% lean and 2% fat	1.00	63.8	4.76	29.7	6.0	0	173	727
225	Fillet, raw, lean and fat	Calculated from 97% lean and 3% fat	1.00	73.6	3.52	22.0	6.5	0	147	615
226	slices, grilled, lean	10 samples	1.00	61.8	5.38	33.6	4.0	0	170	719
227	-, lean and fat	Calculated from 98% lean and 2% fat	1.00	61.1	5.31	33.2	5.0	0	178	749
228	strips, stir-fried, lean	10 samples	1.00	59.6	5.14	32.1	5.9	0	182	764
229	Hand, shoulder joint, raw, lean	10 samples	1.00	74.4	3.33	20.8	4.2	0	121	509
230	-, lean and fat	Calculated from 73% lean and 26% fat	1.00	63.8	2.90	18.1	16.2	0	218	907
231	pressure cooked, lean	10 samples	1.00	61.4	4.83	30.2	7.9	0	192	806
232	-, lean and fat	Calculated from 66% lean and 33% fat	1.00	54.9	3.98	24.9	18.8	0	269	1119
233	roasted, lean	10 samples	1.00	60.0	4.93	30.8	8.8	0	202	849
234	-, lean and fat	Calculated from 71% lean and 27% fat	1.00	55.9	4.40	27.5	16.5	0	259	1078
235	Leg joint, raw, lean	10 samples of a mixture of leg and knuckle joints	1.00	74.9	3.47	21.7	2.2	0	107	450
236	-, lean and fat	Calculated from 79% lean and 21% fat	1.00	64.4	3.04	19.0	15.2	0	213	885
237	-, -, weighed with bone	Calculated from no. 236	0.87	56.0	2.64	16.5	13.2	0	185	769
238	microwaved, lean	10 samples	1.00	61.0	5.38	33.6	4.8	0	178	749
239	-, lean and fat	Calculated from 78% lean and 22% fat	1.00	55.2	4.69	29.3	15.0	0	252	1053
240	roasted medium, lean	10 samples	1.00	61.1	5.28	33.0	5.5	0	182	765
241	-, lean and fat	Calculated from 83% lean and 17% fat	1.00	58.3	4.94	30.9	10.2	0	215	903

No. 18-	Food	Starch g	Total sugars g	Dietary fibre		Fatty acids			Total trans g	Cholesterol mg
				Southgate method g	Englyst method g	*cis & trans*				
						Satd g	Mono- unsatd g	Poly- unsatd g		
221	**Diced**, kebabs, *grilled*, lean	0	0	0	0	1.6	1.8	0.8	Tr	90
222	-, lean and fat	0	0	0	0	2.2	2.4	1.0	0.1	93
223	*slow-cooked*, lean	0	0	0	0	1.6	1.9	1.3	Tr	92
224	-, lean and fat	0	0	0	0	1.8	2.2	1.5	Tr	93
225	**Fillet**, *raw*, lean and fat	0	0	0	0	2.3	2.3	1.3	Tr	63
226	**slices**, *grilled*, lean	0	0	0	0	1.5	1.4	0.8	Tr	89
227	-, lean and fat	0	0	0	0	1.9	1.8	1.0	Tr	89
228	**strips**, *stir-fried*, lean	0	0	0	0	1.3	1.8	2.2	Tr	90
229	**Hand, shoulder joint**, *raw*, lean	0	0	0	0	1.4	1.6	0.8	Tr	64
230	-, lean and fat	0	0	0	0	5.3	6.4	3.1	0.1	68
231	*pressure cooked*, lean	0	0	0	0	2.8	3.1	1.2	Tr	93
232	-, lean and fat	0	0	0	0	6.5	7.6	3.1	0.1	96
233	*roasted*, lean	0	0	0	0	3.0	3.6	1.4	Tr	105
234	-, lean and fat	0	0	0	0	5.7	6.9	2.7	0.1	105
235	**Leg joint**, *raw*, lean	0	0	0	0	0.7	0.9	0.4	Tr	64
236	-, lean and fat	0	0	0	0	5.1	6.4	2.5	0.1	63
237	-, -, weighed with bone	0	0	0	0	4.4	5.6	2.2	0.1	55
238	*microwaved*, lean	0	0	0	0	1.7	2.0	0.7	Tr	105
239	-, lean and fat	0	0	0	0	5.3	6.5	2.1	0.1	100
240	*roasted medium*, lean	0	0	0	0	1.9	2.3	0.7	Tr	100
241	-, lean and fat	0	0	0	0	3.6	4.4	1.4	Tr	100

No. 18-	Food	mg										μg	
		Na	K	Ca	Mg	P	Fe	Cu	Zn	Cl	Mn	Se	I
221	**diced**, kebabs, *grilled*, lean	80	510	9	34	310	1.2	0.12	3.4	68	0.04	21	5
222	-, lean and fat	81	500	9	33	310	1.2	0.12	3.3	69	0.04	21	5
223	*slow-cooked*, lean	41	230	12	20	180	1.1	0.13	3.5	60	0.02	19	5
224	-, lean and fat	42	240	12	20	180	1.1	0.13	3.5	60	0.02	19	5
225	**Fillet**, *raw*, lean and fat	53	400	5	24	230	0.7	0.08	1.6	54	0.02	14	3
226	**slices**, *grilled*, lean	67	520	9	35	320	1.3	0.14	2.7	67	0.02	21	5
227	-, lean and fat	67	520	9	35	310	1.3	0.14	2.7	67	0.02	21	5
228	**strips**, *stir-fried*, lean	71	540	8	35	320	1.4	0.14	2.6	70	0.02	20	3
229	**Hand, shoulder joint**, *raw*, lean	73	370	6	23	210	0.9	0.08	2.6	49	0.02	13	5
230	-, lean and fat	67	320	7	20	180	0.8	0.07	2.1	50	0.01	11	5
231	*pressure cooked*, lean	61	320	10	24	220	1.4	0.14	4.2	60	0.02	19	5
232	-, lean and fat	64	290	11	20	190	1.1	0.11	3.1	63	0.01	16	5
233	*roasted*, lean	78	380	7	26	240	1.3	0.12	4.3	75	0.02	18	7
234	-, lean and fat	77	350	7	24	220	1.2	0.11	3.7	74	0.02	16	7
235	**Leg joint**, *raw*, lean	65	380	6	24	210	0.8	0.02	2.2	52	Tr	14	5
236	-, lean and fat	60	330	6	21	180	0.7	0.02	1.9	50	Tr	12	5
237	-, -, weighed with bone	52	290	5	18	160	0.6	0.02	1.6	43	Tr	10	4
238	*microwaved*, lean	63	360	13	27	240	1.1	0.06	3.1	67	Tr	21	5
239	-, lean and fat	62	330	12	24	220	1.0	0.06	2.6	65	Tr	18	5
240	*roasted medium*, lean	69	400	10	27	250	1.1	0.06	3.2	67	Tr	21	3
241	-, lean and fat	70	380	10	26	240	1.0	0.06	2.9	67	Tr	20	3

Pork continued

No. 18-	Food	Retinol µg	Carotene µg	Vitamin D µg	Vitamin E mg	Thiamin mg	Ribo-flavin mg	Niacin mg	Trypt 60 mg	Vitamin B6 mg	Vitamin B12 µg	Folate µg	Panto-thenate mg	Biotin µg	Vitamin C mg
221	**Diced**, kebabs, *grilled*, lean	Tr	Tr	0.8	0.02	1.11	0.37	9.8	7.1	0.18	1	1	2.20	5	0
222	-, lean and fat	Tr	Tr	0.9	0.02	1.08	0.37	9.6	6.9	0.18	1	1	2.16	5	0
223	*slow-cooked*, lean	Tr	Tr	0.7	0.03	0.52	0.25	4.5	6.2	0.27	1	1	1.94	5	0
224	-, lean and fat	Tr	Tr	0.7	0.03	0.52	0.25	4.5	6.1	0.27	1	1	1.92	5	0
225	**Fillet**, *raw*, lean and fat	Tr	Tr	0.5	Tr	1.16	0.32	6.1	4.5	0.50	1	1	1.49	1	0
226	**slices**, *grilled*, lean	Tr	Tr	0.8	0.05	1.58	0.28	10.5	7.0	0.66	1	4	2.17	5	0
227	-, lean and fat	Tr	Tr	0.8	0.05	1.56	0.28	10.4	6.9	0.65	1	4	2.15	5	0
228	**strips**, *stir-fried*, lean	Tr	Tr	0.8	0.19	1.53	0.41	10.1	6.6	0.78	1	4	2.20	5	0
229	**Hand, shoulder joint**, *raw*, lean	Tr	Tr	0.5	0.01	0.95	0.26	7.8	4.4	0.52	1	3	1.45	2	0
230	-, lean and fat	Tr	Tr	0.8	0.02	0.76	0.23	6.4	3.5	0.41	1	3	1.24	3	0
231	*pressure cooked*, lean	Tr	Tr	0.7	0.01	0.87	0.30	1.5	6.3	0.74	1	2	1.95	5	0
232	-, lean and fat	Tr	Tr	1.2	0.02	0.68	0.26	2.2	4.7	0.54	1	2	1.58	6	0
233	*roasted*, lean	Tr	Tr	0.7	0.02	0.99	0.34	3.7	6.6	0.40	1	1	2.04	4	0
234	-, lean and fat	Tr	Tr	1.0	0.02	0.87	0.30	3.6	5.6	0.35	1	1	1.80	5	0
235	**Leg joint**, *raw*, lean	Tr	Tr	0.5	0.08	0.81	0.20	6.8	4.6	0.50	1	1	1.53	2	0
236	-, lean and fat	Tr	Tr	0.7	0.07	0.68	0.18	5.8	3.9	0.42	1	1	1.32	3	0
237	-, -, weighed with bone	Tr	Tr	0.6	0.06	0.59	0.16	5.0	3.3	0.36	1	1	1.13	2	0
238	*microwaved*, lean	Tr	Tr	0.8	0.02	0.74	0.21	5.9	7.0	0.44	1	1	2.17	5	0
239	-, lean and fat	Tr	Tr	1.0	0.02	0.64	0.20	5.3	5.8	0.38	1	1	1.88	5	0
240	*roasted medium*, lean	Tr	Tr	0.6	0.02	0.73	0.25	9.7	6.7	0.50	1	4	2.90	5	0
241	-, lean and fat	Tr	Tr	0.8	0.03	0.71	0.24	9.2	6.1	0.47	1	4	2.67	5	0

Pork *continued*

Composition of food per 100g

No. 18-	Food	Description and main data sources	Edible Proportion	Water g	Total Nitrogen g	Protein g	Fat g	Carbo-hydrate g	Energy value kcal	Energy value kJ
242	**Leg joint**, *roasted well-done*, lean	10 samples	1.00	60.4	5.55	34.7	5.1	0	185	779
243	-, lean and fat	Calculated from 79% lean and 21% fat	1.00	54.7	4.82	30.1	15.4	0	259	1082
244	frozen, *roasted*, lean	4 samples of a mixture of leg joints from fillet and shank	1.00	60.0	5.18	32.4	7.0	0	193	810
245	-, lean and fat	Calculated from 84% lean and 16% fat	1.00	57.6	4.91	30.7	11.2	0	224	936
246	**Loin chops**, *raw*, lean and fat	Calculated from 70% lean and 30% fat	1.00	59.8	2.98	18.6	21.7	0	270	1119
247	-, -, weighed with bone	Calculated from no. 246	0.84	50.2	2.51	15.7	18.2	0	227	940
248	*barbecued*, lean	19 samples of a mixture of loin and pork chops	1.00	61.0	4.98	31.1	6.8	0	186	780
249	-, lean and fat	Calculated from 82% lean and 18% fat	1.00	55.0	4.53	28.3	15.8	0	255	1066
250	frozen, *grilled*, lean	34 samples	1.00	57.3	5.26	32.9	9.0	0	213	892
251	*grilled*, lean	22 samples of a mixture of loin and pork chops	1.00	61.2	5.06	31.6	6.4	0	184	774
252	-, lean and fat	Calculated from 80% lean and 20% fat	1.00	54.6	4.64	29.0	15.7	0	257	1074
253	*microwaved*, lean	18 samples, mixture of loin and pork chops	1.00	59.5	5.44	34.0	6.1	0	191	804
254	-, lean and fat	Calculated from 82% lean and 18% fat	1.00	55.4	4.83	30.2	14.1	0	248	1035
255	*roasted*, lean	16 samples, mixture of loin and pork chops	1.00	52.7	6.00	37.5	10.1	0	241	1011
256	-, lean and fat	Calculated from 78% lean and 22% fat	1.00	49.1	5.10	31.9	19.3	0	301	1256
257	-, -, weighed with bone	Calculated from no. 256	0.76	37.3	3.88	24.2	14.7	0	229	955
258	**Loin joint**, *raw*, lean and fat	Calculated from 75% lean and 25% fat	1.00	61.9	3.09	19.3	18.8	0	246	1024
259	-, -, weighed with bone	Calculated from no. 258	0.81	50.1	2.50	15.6	15.2	0	199	828
260	*pot-roasted*, lean	10 samples	1.00	58.1	4.96	31.0	8.1	0	197	827
261	-, lean and fat	Calculated from 74% lean and 26% fat	1.00	50.6	3.93	24.6	23.0	0	305	1269

Pork *continued*

No. 18-	Food	Starch g	Total sugars g	Dietary fibre Southgate method g	Englyst method g	Satd g	Fatty acids cis & trans Mono-unsatd g	Poly-unsatd g	Total trans g	Cholesterol mg
242	**Leg joint,** *roasted well-done,* lean	0	0	0	0	1.8	2.1	0.7	Tr	105
243	-, lean and fat	0	0	0	0	5.4	6.6	2.2	0.1	100
244	frozen, *roasted,* lean	0	0	0	0	2.4	2.9	1.0	Tr	98
245	-, lean and fat	0	0	0	0	4.0	4.8	1.6	0.1	98
246	**Loin chops,** *raw,* lean and fat	0	0	0	0	8.0	8.5	3.6	0.1	61
247	-, -, weighed with bone	0	0	0	0	6.7	7.1	3.0	0.1	51
248	*barbecued,* lean	0	0	0	0	2.4	2.6	1.1	Tr	82
249	-, lean and fat	0	0	0	0	5.7	6.3	2.6	0.1	87
250	frozen, *grilled,* lean	0	0	0	0	3.1	3.6	1.4	Tr	87
251	*grilled,* lean	0	0	0	0	2.2	2.6	1.0	Tr	75
252	-, lean and fat	0	0	0	0	5.6	6.5	2.5	0.1	86
253	*microwaved,* lean	0	0	0	0	2.1	2.4	1.0	Tr	105
254	-, lean and fat	0	0	0	0	4.9	5.7	2.5	0.1	100
255	*roasted,* lean	0	0	0	0	3.7	4.0	1.5	0.1	115
256	-, lean and fat	0	0	0	0	7.0	7.8	3.1	0.1	110
257	-, -, weighed with bone	0	0	0	0	5.3	6.5	2.6	0.1	92
258	**Loin joint,** *raw,* lean and fat	0	0	0	0	6.9	7.3	3.1	0.1	60
259	-, -, weighed with bone	0	0	0	0	5.6	5.9	2.5	0.1	49
260	*pot-roasted,* lean	0	0	0	0	2.9	3.2	1.2	Tr	95
261	-, lean and fat	0	0	0	0	8.2	9.3	3.8	0.1	93

Pork *continued*

Inorganic constituents per 100g food

No. 18-	Food	Na	K	Ca	Mg	P	Fe	Cu	Zn	Cl	Mn	Se	I
						(mg)						(µg)	
242	**Leg joint**, *roasted well-done*, lean	74	410	9	28	260	1.2	0.06	3.3	69	Tr	22	5
243	-, lean and fat	71	360	9	25	230	1.1	0.06	2.8	67	Tr	19	5
244	frozen, *roasted*, lean	77	410	7	28	260	1.1	0.08	2.8	64	0.01	20	5
245	-, lean and fat	76	390	7	27	240	1.0	0.08	2.6	64	0.01	19	5
246	**Loin chops**, *raw*, lean and fat	53	300	10	19	170	0.4	0.06	1.3	56	0.01	11	8
247	-, -, weighed with bone	45	250	9	16	140	0.4	0.05	1.1	47	0.01	9	7
248	*barbecued*, lean	67	430	23	29	260	0.8	0.08	2.6	62	0.02	19	5
249	-, lean and fat	68	400	21	26	240	0.8	0.08	2.3	64	0.02	17	5
250	frozen, *grilled*, lean	58	420	27	29	260	0.7	0.06	2.4	65	0.01	21	5
251	*grilled*, lean	66	410	14	28	250	0.7	0.08	2.4	70	0.02	18	3
252	-, lean and fat	70	390	14	26	230	0.7	0.08	2.2	73	0.02	17	3
253	*microwaved*, lean	58	360	21	27	240	0.8	0.08	2.7	67	0.02	21	5
254	-, lean and fat	58	330	19	24	220	0.7	0.07	2.4	65	0.02	19	5
255	*roasted*, lean	70	410	22	29	260	0.9	0.10	2.9	74	0.02	23	5
256	-, lean and fat	68	360	19	25	230	0.8	0.09	2.4	70	0.02	20	5
257	-, -, weighed with bone	51	270	15	19	170	0.6	0.07	1.8	54	0.01	15	4
258	**Loin joint**, *raw*, lean and fat	55	310	10	20	180	0.5	0.06	1.4	55	0.02	11	9
259	-, -, weighed with bone	44	250	8	16	140	0.4	0.05	1.1	45	0.01	9	7
260	*pot-roasted*, lean	54	330	11	24	220	0.9	0.08	3.0	62	0.02	19	5
261	-, lean and fat	56	290	11	20	180	0.8	0.07	2.2	61	0.01	15	5

No. Food 18-	Retinol µg	Carotene µg	Vitamin D µg	Vitamin E mg	Thiamin mg	Riboflavin mg	Niacin mg	Trypt 60 mg	Vitamin B_6 mg	Vitamin B_{12} µg	Folate µg	Pantothenate mg	Biotin µg	Vitamin C mg
242 **Leg joint**, *roasted well-done,* lean	Tr	Tr	0.8	0.04	0.80	0.28	7.9	7.2	0.43	1	1	2.24	5	0
243 -, lean and fat	Tr	Tr	1.0	0.04	0.69	0.25	6.9	6.0	0.37	1	1	1.93	5	0
244 frozen, *roasted,* lean	Tr	Tr	0.8	0.04	1.08	0.26	8.2	6.7	0.34	1	4	2.09	5	0
245 -, lean and fat	Tr	Tr	0.9	0.04	1.01	0.25	7.8	6.2	0.32	1	4	1.97	5	0
246 **Loin chops**, *raw,* lean and fat	Tr	Tr	0.9	0.11	0.81	0.18	4.9	3.3	0.62	1	1	0.97	3	0
247 -, -, weighed with bone	Tr	Tr	0.8	0.09	0.68	0.15	4.2	2.8	0.52	1	1	0.81	2	0
248 *barbecued,* lean	Tr	Tr	0.7	0.01	1.18	0.17	9.6	6.4	0.35	1	1	2.01	5	0
249 -, lean and fat	Tr	Tr	1.0	0.02	1.03	0.17	8.6	5.6	0.32	1	1	1.82	6	0
250 frozen, *grilled,* lean	Tr	Tr	0.8	0.03	1.12	0.19	8.5	6.8	0.38	1	4	2.12	5	0
251 *grilled,* lean	Tr	Tr	0.8	0.01	0.78	0.16	9.1	6.2	0.56	1	7	1.22	4	0
252 -, lean and fat	Tr	Tr	1.1	0.02	0.70	0.17	8.2	5.3	0.49	1	6	1.20	5	0
253 *microwaved,* lean	Tr	Tr	0.8	0.03	1.06	0.17	7.9	7.0	0.42	1	4	2.19	5	0
254 -, lean and fat	Tr	Tr	1.0	0.03	0.92	0.17	7.0	6.0	0.37	1	4	1.93	5	0
255 *roasted,* lean	Tr	Tr	0.8	0.03	0.91	0.16	9.9	7.8	0.50	1	2	2.42	6	0
256 -, lean and fat	Tr	Tr	1.1	0.03	0.77	0.16	8.4	6.3	0.42	1	2	2.05	6	0
257 -, -, weighed with bone	Tr	Tr	0.8	0.02	0.58	0.12	6.4	4.8	0.32	1	2	1.56	5	0
258 **Loin joint**, *raw,* lean and fat	Tr	Tr	0.9	0.12	0.87	0.18	5.2	3.5	0.66	1	1	1.00	3	0
259 -, -, weighed with bone	Tr	Tr	0.7	0.10	0.70	0.15	4.2	2.8	0.53	1	1	0.81	2	0
260 *pot-roasted,* lean	Tr	Tr	0.7	0.02	0.69	0.21	5.2	6.4	0.35	1	6	2.00	5	0
261 -, lean and fat	Tr	Tr	1.1	0.03	0.55	0.19	4.5	4.7	0.28	1	5	1.57	6	0

No. 18-	Food	Description and main data sources	Edible Proportion	Water g	Total Nitrogen g	Protein g	Fat g	Carbohydrate g	Energy value kcal	kJ
262	**Loin joint**, *roasted*, lean	10 samples	1.00	61.4	4.82	30.1	6.8	0	182	763
263	-, lean and fat	Calculated from 74% lean and 26% fat	1.00	56.0	4.20	26.3	16.4	0	253	1054
264	**Loin steaks**, *raw*, lean and fat	10 samples	1.00	63.9	3.19	19.9	16.1	0	225	934
265	*fried*, lean	22 samples of a mixture of loin steaks, boneless chops and noisettes	1.00	60.2	5.04	31.5	7.2	0	191	802
266	-, lean and fat	Calculated from 76% lean and 23% fat	1.00	52.2	4.40	27.5	18.4	0	276	1148
267	**Mince**, *raw*	10 samples	1.00	70.4[a]	3.07	19.2	9.7[b]	0	164	685
268	*stewed*	10 samples	1.00	64.7	3.90	24.4	10.4	0	191	800
269	**Spare rib chops**, *raw*, lean and fat	From shoulder; Calculated from 93% lean and 7% fat	1.00	69.8	2.96	18.5	12.4	0	186	773
270	-, -, weighed with bone	Calculated from no. 269	0.79	55.1	2.34	14.6	9.8	0	147	611
271	*braised*, lean	26 samples	1.00	58.8	4.88	30.5	10.1	0	213	892
272	-, lean and fat	Calculated from 87% lean and 13% fat	1.00	55.7	4.49	28.1	15.0	0	247	1033
273	-, -, weighed with bone	Calculated from no. 272	0.77	42.9	3.46	21.6	11.5	0	190	793
274	**Spare rib joint**, *raw*, lean and fat	From shoulder; Calculated from 74% lean and 26% fat	1.00	66.8	2.82	17.6	16.4	0	218	906
275	-, -, weighed with bone	Calculated from no. 274	0.74	49.4	2.09	13.1	12.1	0	161	670
276	*pot-roasted*, lean	10 samples	1.00	60.0	4.80	30.0	9.0	0	201	843
277	-, lean and fat	Calculated from 67% lean and 32% fat	1.00	55.9	3.91	24.4	18.0	0	260	1081

[a] Water ranged from 64.3g to 73.2g per 100g

[b] Fat ranged from 5.4g to 19.5g per 100g

Pork continued

No. Food 18-	Starch g	Total sugars g	Dietary fibre Southgate method g	Dietary fibre Englyst method g	Fatty acids cis & trans Satd g	Mono-unsatd g	Poly-unsatd g	Total trans g	Cholesterol mg
262 **Loin joint**, *roasted*, lean	0	0	0	0	2.4	2.6	1.1	Tr	78
263 -, lean and fat	0	0	0	0	5.9	6.5	2.8	0.1	80
264 **Loin steaks**, *raw*, lean and fat	0	0	0	0	5.9	6.3	2.7	0.1	59
265 *fried*, lean	0	0	0	0	2.3	2.8	1.4	0.1	86
266 -, lean and fat	0	0	0	0	6.0	7.2	3.6	0.1	90
267 **Mince**, *raw*	0	0	0	0	3.6	3.8	1.6	Tr	66
268 *stewed*	0	0	0	0	3.9	4.2	1.6	0.1	81
269 **Spare rib chops**, *raw*, lean and fat	0	0	0	0	4.5	4.7	2.3	0.1	67
270 -, -, weighed with bone	0	0	0	0	3.5	3.7	1.8	0.1	53
271 *braised*, lean	0	0	0	0	3.6	3.8	1.7	0.1	98
272 -, lean and fat	0	0	0	0	5.5	5.8	2.6	0.1	97
273 -, -, weighed with bone	0	0	0	0	4.2	4.5	2.0	0.1	75
274 **Spare rib joint**, *raw*, lean and fat	0	0	0	0	6.0	6.2	3.0	0.1	68
275 -, -, weighed with bone	0	0	0	0	4.4	4.6	2.2	0.1	50
276 *pot-roasted*, lean	0	0	0	0	3.2	3.4	1.5	0.1	92
277 -, lean and fat	0	0	0	0	6.7	6.9	3.2	0.1	92

Inorganic constituents per 100g food

No. 18-	Food	Na	K	Ca	Mg	P	Fe	Cu	Zn	Cl	Mn	Se	I
						mg						µg	
262	**Loin joint**, *roasted*, lean	61	370	12	25	230	0.8	0.06	2.8	60	Tr	19	5
263	-, lean and fat	61	330	12	22	200	0.7	0.06	2.4	60	Tr	17	5
264	**Loin steaks**, *raw*, lean and fat	56	320	10	21	180	0.5	0.06	1.5	55	0.02	11	9
265	*fried*, lean	58	430	6	28	260	0.7	0.08	2.3	63	0.02	20	5
266	-, lean and fat	61	390	7	25	230	0.7	0.07	2.0	65	0.02	18	5
267	**Mince**, *raw*	66	390	7	20	190	0.9	0.06	2.4	67	0.01	12	5
268	*stewed*	61	320	13	21	200	1.4	0.10	3.5	61	0.02	15	3
269	**Spare rib chops**, *raw*, lean and fat	62	340	8	21	190	0.8	0.07	2.5	48	Tr	12	5
270	-, -, weighed with bone	49	270	6	16	150	0.6	0.06	2.0	38	Tr	9	4
271	*braised*, lean	59	300	32	23	210	1.5	0.12	4.9	54	0.02	20	11
272	-, lean and fat	59	290	29	22	200	1.4	0.11	4.4	55	0.02	18	11
273	-, -, weighed with bone	44	220	22	16	150	1.0	0.08	3.3	41	0.01	14	8
274	**Spare rib joint**, *raw*, lean and fat	61	320	8	20	180	0.8	0.07	2.3	48	Tr	11	5
275	-, -, weighed with bone	45	240	6	15	140	0.6	0.05	1.7	36	Tr	8	4
276	*pot-roasted*, lean	65	320	8	24	220	1.3	0.12	4.5	60	0.02	19	5
277	-, lean and fat	64	280	9	20	190	1.0	0.10	3.3	60	0.01	15	5

No. 18-	Food	Retinol µg	Carotene µg	Vitamin D µg	Vitamin E mg	Thiamin mg	Ribo- flavin mg	Niacin mg	Trypt 60 mg	Vitamin B6 mg	Vitamin B12 µg	Folate µg	Panto- thenate mg	Biotin µg	Vitamin C mg
262	**Loin joint**, *roasted*, lean	Tr	Tr	0.7	0.03	0.78	0.20	6.5	6.2	0.37	1	2	1.94	5	0
263	-, lean and fat	Tr	Tr	1.0	0.03	0.67	0.19	5.8	5.2	0.32	1	2	1.68	5	0
264	**Loin steaks**, *raw*, lean and fat	Tr	Tr	0.9	0.12	0.92	0.18	5.4	3.6	0.69	1	1	1.02	3	0
265	*fried*, lean	Tr	Tr	0.8	0.10	0.82	0.17	9.5	6.5	0.68	1	11	2.03	5	0
266	-, lean and fat	Tr	Tr	1.1	0.09	0.70	0.17	8.1	5.3	0.56	1	9	1.76	6	0
267	**Mince**, *raw*	Tr	Tr	0.8	0.09	0.96	0.22	6.5	3.5	0.20	1	2	1.31	1	0
268	*stewed*	Tr	Tr	0.9	0.02	0.97	0.22	6.2	4.4	0.23	1	2	1.58	4	0
269	**Spare rib chops**, *raw*, lean and fat	Tr	Tr	0.6	0.01	0.76	0.28	5.0	3.7	0.40	1	10	1.25	2	0
270	-, -, weighed with bone	Tr	Tr	0.5	0.01	0.60	0.22	3.9	2.9	0.31	1	8	0.99	1	0
271	*braised*, lean	Tr	Tr	0.4	0.04	0.77	0.29	3.6	6.3	0.26	2	1	1.61	6	0
272	-, lean and fat	Tr	Tr	0.6	0.04	0.70	0.27	3.5	5.7	0.24	1	1	1.50	6	0
273	-, -, weighed with bone	Tr	Tr	0.4	0.03	0.53	0.20	2.7	4.3	0.18	1	1	1.12	5	0
274	**Spare rib joint**, *raw*, lean and fat	Tr	Tr	0.8	0.02	0.71	0.26	4.7	3.4	0.37	1	9	1.19	2	0
275	-, -, weighed with bone	Tr	Tr	0.6	0.01	0.52	0.20	3.5	2.5	0.27	1	7	0.88	2	0
276	*pot-roasted*, lean	Tr	Tr	0.7	0.02	0.58	0.29	4.7	6.2	0.40	1	4	1.94	5	0
277	-, lean and fat	Tr	Tr	1.1	0.03	0.48	0.25	4.2	4.6	0.32	1	3	1.56	6	0

No. 18-	Food	Description and main data sources	Edible Proportion	Water g	Total Nitrogen g	Protein g	Fat g	Carbo- hydrate g	Energy value kcal	kJ
278	**Spare ribs**, *raw*, lean and fat	10 samples	1.00	67.3	2.99	18.7	13.4	0	195	814
279	-, -, weighed with bone	Calculated from 41% lean and 6% fat	0.48	32.3	1.44	9.0	6.4	0	94	390
280	*grilled*, lean and fat	10 samples	1.00	50.5	4.67	29.2	19.5	0	292	1218
281	-, -, weighed with bone	Calculated from 39% lean and 6% fat	0.46	22.7	2.10	13.1	8.8	0	132	548
282	**Spare rib steaks**, *raw*, lean and fat	From shoulder; Calculated from 86% lean and 13% fat	1.00	70.4	2.99	18.7	10.4	0	168	703
283	**Steaks**, *raw*, lean	15 samples of a mixture of pork and leg steaks	1.00	73.9	3.58	22.4	3.4	0	120	507
284	-, lean and fat	Calculated from 89% lean and 11% fat	1.00	69.6	3.36	21.0	9.4	0	169	705
285	*grilled*, lean	19 samples of a mixture of pork and leg steaks	1.00	61.5	5.42	33.9	3.7	0	169	713
286	-, lean and fat	Calculated from 92% lean and 8% fat	1.00	59.1	5.19	32.4	7.6	0	198	832
287	*stewed*, lean	15 samples of a mixture of pork and leg steaks	1.00	61.7	5.38	33.6	4.6	0	176	741
288	-, lean and fat	Calculated from 89% lean and 9% fat	1.00	58.6	4.96	31.0	8.3	0	199	834

Pork *continued*

No. 18-	Food	Starch g	Total sugars g	Dietary fibre Southgate method g	Englyst method g	Fatty acids Satd g	cis & trans Mono-unsatd g	Poly-unsatd g	Total trans g	Cholesterol mg
278	**Spare ribs**, *raw*, lean and fat	0	0	0	0	5.2	5.3	1.6	0.1	66
279	-, -, weighed with bone	0	0	0	0	2.5	2.5	0.8	Tr	31
280	*grilled*, lean and fat	0	0	0	0	7.5	8.2	2.4	0.1	115
281	-, -, weighed with bone	0	0	0	0	3.4	3.9	1.1	Tr	55
282	**Spare rib steaks**, *raw*, lean and fat	0	0	0	0	3.7	3.9	1.9	0.1	66
283	**Steaks**, *raw*, lean	0	0	0	0	1.2	1.3	0.6	Tr	62
284	-, lean and fat	0	0	0	0	3.3	3.8	1.6	0.1	63
285	*grilled*, lean	0	0	0	0	1.3	1.4	0.6	Tr	99
286	-, lean and fat	0	0	0	0	2.7	3.0	1.2	Tr	100
287	*stewed*, lean	0	0	0	0	1.3	1.6	1.2	Tr	105
288	-, lean and fat	0	0	0	0	2.4	3.0	2.1	0.1	100

Pork *continued*

Inorganic constituents per 100g food

No. 18-	Food	Na	K	Ca	Mg	P	Fe	Cu	Zn	Cl	Mn	Se	I
							mg					µg	
278	**Spare ribs**, *raw*, lean and fat	98	290	15	18	160	0.8	0.04	2.5	65	Tr	12	5
279	-, -, weighed with bone	46	140	7	8	73	0.4	0.02	1.2	31	Tr	6	2
280	*grilled*, lean and fat	140	390	30	27	240	1.4	0.08	4.0	130	Tr	18	3
281	-, -, weighed with bone	63	180	14	12	110	0.6	0.04	1.8	56	Tr	8	1
282	**Spare rib steaks**, *raw*, lean and fat	62	350	8	21	200	0.8	0.07	2.6	47	Tr	12	5
283	**Steaks**, *raw*, lean	60	390	6	24	220	0.7	0.02	1.8	52	Tr	14	5
284	-, lean and fat	58	360	6	22	210	0.7	0.02	1.7	51	Tr	13	5
285	*grilled*, lean	76	480	8	33	300	1.1	0.10	2.9	67	0.02	21	5
286	-, lean and fat	76	460	8	32	280	1.1	0.10	2.7	68	0.02	20	5
287	*stewed*, lean	46	290	12	25	220	1.3	0.16	3.5	67	0.02	21	5
288	-, lean and fat	46	280	12	23	210	1.2	0.15	3.2	65	0.02	19	5

No. Food 18-	Retinol µg	Carotene µg	Vitamin D µg	Vitamin E mg	Thiamin mg	Ribo-flavin mg	Niacin mg	Trypt 60 mg	Vitamin B6 mg	Vitamin B12 µg	Folate µg	Panto-thenate mg	Biotin µg	Vitamin C mg
278 **Spare ribs**, *raw*, lean and fat	Tr	Tr	0.8	0.01	0.76	0.18	5.0	3.4	0.33	1	2	1.53	1	0
279 -, -, weighed with bone	Tr	Tr	0.4	Tr	0.36	0.08	2.4	1.6	0.16	1	1	0.72	Tr	0
280 *grilled*, lean and fat	Tr	Tr	1.2	0.02	0.73	0.37	7.1	5.2	0.50	1	4	1.88	4	0
281 -, -, weighed with bone	Tr	Tr	0.5	0.01	0.33	0.17	3.2	2.4	0.23	Tr	2	0.85	2	0
282 **Spare rib steaks**, *raw*, lean and fat	Tr	Tr	0.6	0.01	0.78	0.29	5.0	3.7	0.41	1	11	1.27	2	0
283 **Steaks**, *raw*, lean	Tr	Tr	0.5	0.05	0.93	0.23	7.4	4.6	0.59	1	1	1.53	2	0
284 -, lean and fat	Tr	Tr	0.6	0.05	0.85	0.22	6.8	4.2	0.54	1	1	1.42	2	0
285 *grilled*, lean	Tr	Tr	0.8	0.02	1.55	0.28	9.5	7.0	0.72	1	9	2.19	5	0
286 -, lean and fat	Tr	Tr	0.9	0.02	1.45	0.27	9.1	6.6	0.68	1	8	2.09	5	0
287 *stewed*, lean	Tr	Tr	0.8	0.02	0.81	0.25	5.9	7.0	0.42	1	1	2.17	5	0
288 -, lean and fat	Tr	Tr	0.8	0.02	0.75	0.24	5.5	6.3	0.39	1	1	2.00	5	0

No. 18-	Food	Description and main data sources	Edible Proportion	Water g	Total Nitrogen g	Protein g	Fat g	Carbohydrate g	Energy value kcal	kJ
Raw										
289	**Dark meat**, *raw*	31 samples	1.00	75.8	3.34	20.9	2.8	0	109	459
290	**Light meat**, *raw*	31 samples	1.00	74.2	3.84	24.0	1.1	0	106	449
291	**Meat**, average, *raw*	Calculated from 44% light meat and 56% dark meat	1.00	75.1	3.57	22.3	2.1	0	108	457
292	**Skin**, *raw*	25 samples	1.00	42.7	1.55	9.7	48.3	0	474	1953
293	**Leg quarter**, *raw*, meat and skin	Calculated from 77% dark meat and 23% skin	1.00	68.1	2.93	18.3	13.3	0	193	803
294	- , - , weighed with bone	Calculated from no. 293	0.69	47.0	2.02	12.6	9.2	0	133	555
295	**Wing quarter**, *raw*, meat and skin	Calculated from 65% light meat, 11% dark meat and 24% skin	1.00	67.0	3.25	20.3	12.4	0	193	804
296	- , - , weighed with bone	Calculated from no. 295	0.66	44.2	2.14	13.4	8.2	0	127	531
297	**Whole chicken**, *raw*	Calculated from 33% light meat, 42% dark meat and 25% skin	1.00	66.8	3.06	19.1	13.8	0	201	835
298	- , - , weighed with bone	Calculated from no. 297	0.65	43.4	1.98	12.4	9.0	0	131	544
299	**Whole chicken, corn-fed**, *raw*, dark meat only	6 samples	1.00	72.5	3.12	19.5	7.2	0	143	599
300	- , light meat only	6 samples	1.00	73.4	3.69	23.0	2.8	0	118	496
301	- , meat only	Calculated from 54% light meat and 46% dark meat	1.00	73.0	3.42	21.4	4.8	0	129	541
302	- , skin only	6 samples	1.00	39.8	1.61	10.1	52.7	0	514	2120
303	- , meat and skin	Calculated from 42% light meat, 35% dark meat and 23% skin	1.00	65.3	3.01	18.8	15.9	0	218	908
304	- , - , weighed with bone	Calculated from no. 303	0.65	42.4	1.95	12.2	10.3	0	142	589

Chicken

No. Food 18-	Starch g	Total sugars g	Dietary fibre Southgate method g	Dietary fibre Englyst method g	Fatty acids Satd g	cis & trans Mono-unsatd g	cis & trans Poly-unsatd g	Total trans g	Cholesterol mg
Raw									
289 **Dark meat**, *raw*	0	0	0	0	0.8	1.3	0.6	Tr	105
290 **Light meat**, *raw*	0	0	0	0	0.3	0.5	0.2	Tr	70
291 **Meat**, average, *raw*	0	0	0	0	0.6	1.0	0.4	Tr	90
292 **Skin**, *raw*	0	0	0	0	13.6	23.6	8.1	0.6	135
293 **Leg quarter**, *raw*, meat and skin	0	0	0	0	3.6	6.3	2.5	0.1	110
294 -, -, weighed with bone	0	0	0	0	2.5	4.3	1.7	0.1	77
295 **Wing quarter**, *raw*, meat and skin	0	0	0	0	3.5	5.6	2.3	0.1	89
296 -, -, weighed with bone	0	0	0	0	2.3	3.7	1.5	0.1	59
297 **Whole chicken**, *raw*	0	0	0	0	3.8	6.4	2.6	0.1	100
298 -, -, weighed with bone	0	0	0	0	2.5	4.2	1.7	0.1	66
299 **Whole chicken, corn-fed**, *raw*, dark meat only	0	0	0	0	2.1	3.2	1.5	Tr	100
300 -, light meat only	0	0	0	0	0.8	1.2	0.6	Tr	70
301 -, meat only	0	0	0	0	1.4	2.1	1.0	Tr	85
302 -, skin only	0	0	0	0	15.2	23.8	10.1	0.3	140
303 -, meat and skin	0	0	0	0	4.6	6.9	3.2	0.1	98
304 -, -, weighed with bone	0	0	0	0	3.0	4.5	2.1	0.1	64

Chicken

Inorganic constituents per 100g food

No. 18-	Food	Na	K	Ca	Mg	P	mg Fe	Cu	Zn	Cl	Mn	µg Se	I
Raw													
289	**Dark meat**, *raw*	90	390	7	24	110	0.8	0.02	1.7	110	0.01	14	6
290	**Light meat**, *raw*	60	370	5	29	220	0.5	0.05	0.7	77	0.01	12	6
291	**Meat**, average, *raw*	77	380	6	26	160	0.7	0.03	1.2	95	0.01	13	6
292	**Skin**, *raw*	50	130	8	9	91	0.6	0.10	0.7	57	0.01	9	16
293	**Leg quarter**, *raw*, meat and skin	80	330	7	21	110	0.8	0.04	1.4	100	0.01	13	8
294	-, -, weighed with bone	55	230	5	14	75	0.5	0.03	1.0	70	0.01	9	5
295	**Wing quarter**, *raw*, meat and skin	60	320	6	24	180	0.6	0.06	0.8	75	0.01	12	8
296	-, -, weighed with bone	40	210	4	16	120	0.4	0.04	0.5	50	0.01	8	5
297	**Whole chicken**, *raw*	70	320	7	22	140	0.7	0.05	1.1	85	0.01	12	8
298	-, -, weighed with bone	45	210	4	14	90	0.4	0.03	0.7	55	0.01	8	5
299	**Whole chicken, corn-fed**, *raw*, dark meat only	80	300	7	23	200	0.9	0.05	1.6	83	0.01	14	5
300	-, light meat only	90	330	9	28	230	0.4	0.09	0.8	66	0.01	12	5
301	-, meat only	85	320	8	26	210	0.6	0.07	1.2	75	0.01	13	5
302	-, skin only	50	130	6	9	95	0.6	0.02	0.7	64	0.01	N	5
303	-, meat and skin	80	270	8	22	190	0.6	0.06	1.1	70	0.01	12	5
304	-, -, weighed with bone	50	180	5	14	120	0.3	0.04	0.7	45	0.01	8	3

No. 18-	Food	Retinol µg	Carotene µg	Vitamin D µg	Vitamin E mg	Thiamin mg	Ribo-flavin mg	Niacin mg	Trypt 60 mg	Vitamin B6 mg	Vitamin B12 µg	Folate µg	Panto-thenate mg	Biotin µg	Vitamin C mg
Raw															
289	**Dark meat**, *raw*	20	Tr	0.1	0.17	0.14	0.22	5.6	4.1	0.28	1	9	1.09	3	0
290	**Light meat**, *raw*	Tr	Tr	0.2	0.13	0.14	0.14	10.7	4.7	0.51	Tr	14	1.26	2	0
291	**Meat**, average, *raw*	11	Tr	0.1	0.15	0.14	0.18	7.8	4.3	0.38	Tr	11	1.16	2	0
292	**Skin**, *raw*	64	Tr	1.9	0.15	0.04	0.44	2.7	0.9	0.08	1	4	0.64	2	0
293	**Leg quarter**, *raw*, meat and skin	30	Tr	0.5	0.17	0.12	0.27	4.9	3.4	0.23	1	8	0.99	3	0
294	-, -, weighed with bone	21	Tr	0.4	0.12	0.08	0.19	3.4	2.3	0.16	Tr	5	0.68	2	0
295	**Wing quarter**, *raw*, meat and skin	17	Tr	0.6	0.14	0.12	0.22	8.3	3.7	0.38	Tr	11	1.10	2	0
296	-, -, weighed with bone	11	Tr	0.4	0.09	0.08	0.14	5.5	2.5	0.25	Tr	7	0.73	1	0
297	**Whole chicken**, *raw*	25	Tr	0.6	0.15	0.11	0.25	6.5	3.5	0.30	Tr	9	1.03	2	0
298	-, -, weighed with bone	16	Tr	0.4	0.10	0.07	0.16	4.2	2.3	0.20	Tr	6	0.67	1	0
299	**Whole chicken, corn-fed**, *raw*, dark meat only	31	Tr	0.5	0.05	0.09	0.22	5.9	3.6	0.34	1	6	1.28	3	0
300	-, light meat only	14	Tr	0.5	0.19	0.07	0.11	11.5	4.4	0.77	Tr	10	1.20	2	0
301	-, meat only	22	Tr	0.5	0.13	0.08	0.16	8.9	4.0	0.57	Tr	8	1.24	2	0
302	-, skin only	165	Tr	N	N	N	N	N	N	N	N	N	N	N	0
303	-, meat and skin	55	Tr	0.5	0.13	0.07	0.13	7.5	3.3	0.46	1	7	1.10	2	0
304	-, -, weighed with bone	36	Tr	0.3	0.08	0.04	0.08	4.9	2.2	0.30	Tr	4	0.71	1	0

Chicken *continued*

Composition of food per 100g

No. 18-	Food	Description and main data sources	Edible Proportion	Water g	Total Nitrogen g	Protein g	Fat g	Carbohydrate g	Energy value kcal	kJ
Raw continued										
305	**Poussin**, *raw*, meat and skin	Calculated from 34% light meat, 40% dark meat and 26% skin	1.00	66.8	3.06	19.1	13.9	0	202	839
306	-, -, weighed with bone	Calculated from no. 305	0.58	38.7	1.78	11.1	8.1	0	117	488
Casseroled										
307	**Breast**, *casseroled*, meat only	Calculated from light meat from fresh and frozen chicken	1.00	67.7	4.54	28.4	5.2	0	114	483
308	-, meat and skin	Calculated from 90% light meat and 10% skin from fresh and frozen chicken	1.00	65.8	4.30	26.9	8.5	0	184	772
309	-, -, weighed with bone	Calculated from no. 308	0.90	59.2	3.87	24.2	7.6	0	165	693
310	**Drumsticks**, *casseroled*, meat only	Calculated from dark meat	1.00	66.5	3.89	24.3	9.7	0	185	772
311	-, meat and skin	Calculated from 85% dark meat and 15% skin from fresh and frozen chicken	1.00	63.7	3.57	22.3	14.2	0	217	905
312	-, -, weighed with bone	Calculated from no. 311	0.65	41.4	2.32	14.5	9.2	0	141	587
313	**Leg quarter**, *casseroled*, meat only	Calculated from 25% light meat and 75% dark meat from fresh and frozen chicken	1.00	66.9	4.00	25.0	8.4	0	176	736
314	-, meat and skin	Calculated from 21% light meat, 62% dark meat and 17% skin from fresh and frozen chicken	1.00	63.6	3.66	22.9	13.9	0	217	904
315	-, -, weighed with bone	Calculated from no. 314	0.63	40.1	2.30	14.4	8.8	0	137	570
316	**Skin**, *casseroled*	20 samples	1.00	48.6	2.14	13.4	39.3	0	407	1680

No. 18-	Food	Starch g	Total sugars g	Dietary fibre Southgate method g	Dietary fibre Englyst method g	Fatty acids Satd g	cis & trans Mono-unsatd g	cis & trans Poly-unsatd g	Total trans g	Cholesterol mg
Raw continued										
305	**Poussin**, *raw*, meat and skin	0	0	0	0	3.8	6.4	2.6	0.1	100
306	-, -, weighed with bone	0	0	0	0	2.2	3.7	1.5	0.1	58
Casseroled										
307	**Breast**, *casseroled*, meat only	0	0	0	0	1.5	2.4	1.0	0.1	90
308	-, meat and skin	0	0	0	0	2.4	3.9	1.6	0.1	92
309	-, -, weighed with bone	0	0	0	0	2.1	3.5	1.4	0.1	83
310	**Drumsticks**, *casseroled*, meat only	0	0	0	0	2.6	4.5	1.9	0.1	135
311	-, meat and skin	0	0	0	0	3.8	6.6	2.8	0.1	125
312	-, -, weighed with bone	0	0	0	0	2.6	4.5	1.9	0.1	80
313	**Leg quarter**, *casseroled*, meat only	0	0	0	0	2.3	3.9	1.6	0.1	115
314	-, meat and skin	0	0	0	0	3.8	6.5	2.6	0.1	115
315	-, -, weighed with bone	0	0	0	0	2.4	4.1	1.7	0.1	72
316	**Skin**, *casseroled*	0	0	0	0	11.0	19.1	6.5	0.5	105

Chicken *continued*

Inorganic constituents per 100g food

No. 18-	Food	Na	K	Ca	Mg	P	mg Fe	Cu	Zn	Cl	Mn	µg Se	I
	Raw continued												
305	**Poussin**, *raw*, meat and skin	70	320	7	22	150	0.6	0.05	1.1	85	0.01	12	8
306	-, -, weighed with bone	40	180	4	13	85	0.3	0.03	0.6	50	0.01	7	5
	Casseroled												
307	**Breast**, *casseroled*, meat only	60	270	9	25	210	0.5	0.06	1.1	60	0.01	13	8
308	-, meat and skin	60	260	10	24	200	0.5	0.06	1.1	60	0.01	14	8
309	-, -, weighed with bone	55	240	9	22	180	0.4	0.05	1.0	55	0.01	13	7
310	**Drumsticks**, *casseroled*, meat only	70	220	30	21	180	1.3	0.06	2.2	70	0.02	17	7
311	-, meat and skin	75	200	30	18	170	1.1	0.07	1.9	70	0.02	15	7
312	-, -, weighed with bone	50	130	20	12	110	0.7	0.04	1.2	45	0.01	10	5
313	**Leg quarter**, *casseroled*, meat only	75	220	26	20	190	0.9	0.07	1.8	70	0.02	15	6
314	-, meat and skin	70	210	24	19	180	0.9	0.07	1.7	70	0.02	14	6
315	-, -, weighed with bone	45	130	15	12	110	0.6	0.04	1.1	45	0.01	9	4
316	**Skin**, *casseroled*	55	150	17	14	130	1.0	0.09	1.0	N	0.02	N	N

Chicken continued

No.18-	Food	Retinol µg	Carotene µg	Vitamin D µg	Vitamin E mg	Thiamin mg	Ribo-flavin mg	Niacin mg	Trypt 60 mg	Vitamin B6 mg	Vitamin B12 µg	Folate µg	Panto-thenate mg	Biotin µg	Vitamin C mg
Raw *continued*															
305	**Poussin,** *raw,* meat and skin	24	Tr	0.6	0.15	0.11	0.25	6.6	3.5	0.31	Tr	9	1.03	2	0
306	-, -, weighed with bone	14	Tr	0.3	0.09	0.06	0.14	3.8	2.0	0.18	Tr	5	0.60	1	0
Casseroled															
307	**Breast,** *casseroled,* meat only	Tr	Tr	0.1	0.07	0.06	0.13	8.8	5.6	0.36	Tr	6	1.34	2	0
308	-, meat and skin	6	Tr	0.1	0.07	0.06	0.13	8.3	5.1	0.34	Tr	6	1.27	2	0
309	-, -, weighed with bone	5	Tr	0.1	0.06	0.05	0.12	7.5	4.6	0.31	Tr	5	1.14	2	0
310	**Drumsticks,** *casseroled,* meat only	21	Tr	Tr	0.60	0.06	0.09	5.1	4.8	0.22	Tr	9	1.09	3	0
311	-, meat and skin	27	Tr	0.2	0.35	0.05	0.12	5.0	4.2	0.21	Tr	8	1.08	3	0
312	-, -, weighed with bone	18	Tr	0.1	0.23	0.03	0.08	3.2	2.7	0.14	Tr	5	0.70	2	0
313	**Leg quarter,** *casseroled,* meat only	15	Tr	0.1	0.31	0.05	0.13	6.1	4.9	0.25	Tr	8	1.21	3	0
314	-, meat and skin	23	Tr	0.2	0.28	0.05	0.12	5.7	4.3	0.23	Tr	7	1.11	3	0
315	-, -, weighed with bone	14	Tr	0.1	0.18	0.03	0.08	3.6	2.7	0.14	Tr	4	0.70	2	0
316	**Skin,** *casseroled*	N	Tr	0.7	N	N	N	N	N	N	N	N	N	N	0

Composition of food per 100g

No. 18-	Food	Description and main data sources	Edible Proportion	Water g	Total Nitrogen g	Protein g	Fat g	Carbohydrate g	Energy value kcal	kJ
	Casseroled continued									
317	**Thighs**, *casseroled*, meat and skin	Calculated from 77% dark meat and 23% skin from fresh and frozen chicken	1.00	62.4	3.44	21.5	16.3	0	233	969
318	-, -, weighed with bone	Calculated from no. 317	0.83	51.8	2.85	17.8	13.5	0	193	802
319	diced, *casseroled*, meat only	10 samples of skinless thighs	1.00	65.6	4.10	25.6	8.6	0	180	756
320	**Wing quarter**, *casseroled*, meat only	Calculated from 74% light meat and 26% dark meat from fresh and frozen chicken	1.00	67.7	4.30	26.9	6.3	0	164	690
321	-, meat and skin	Calculated from 60% light meat, 21% dark meat and 19% skin	1.00	64.0	3.90	24.4	12.5	0	210	877
322	-, -, weighed with bone	Calculated from no. 321	0.67	42.9	2.61	16.3	8.4	0	141	588
	Grilled and fried									
323	**Breast**, *grilled without skin*, meat only	10 samples	1.00	66.6	5.11	32.0	2.2	0	148	626
324	*grilled with skin*, meat only	17 samples	1.00	68.0	4.76	29.8	3.1	0	147	621
325	-, meat and skin	Calculated from 92% light meat and 8% skin from fresh and frozen chicken	1.00	64.8	4.63	28.9	6.4	0	173	728
326	**Breast**, strips, *stir-fried*	10 samples	1.00	65.9	4.76	29.7	4.6	0	161	677
327	**Portions**, *deep-fried*, meat and skin	10 samples of wing quarters, leg quarters and chicken halves from fish and chip shops	1.00	55.5	4.30	26.9	16.8	0	259	1079
328	-, -, weighed with bone	Calculated from no. 327	0.69	38.2	2.98	18.6	11.6	0	179	745

No. Food 18-	Starch g	Total sugars g	Dietary fibre Southgate method g	Dietary fibre Englyst method g	Fatty acids cis & trans Satd g	Fatty acids cis & trans Mono-unsatd g	Fatty acids cis & trans Poly-unsatd g	Total trans g	Cholesterol mg
Casseroled *continued*									
317 **Thighs**, *casseroled*, meat and skin	0	0	0	0	4.4	7.6	3.1	0.2	120
318 - , weighed with bone	0	0	0	0	3.7	6.3	2.6	0.1	100
319 diced, *casseroled*, meat only	0	0	0	0	2.4	3.8	1.9	Tr	130
320 **Wing quarter**, *casseroled*, meat only	0	0	0	0	1.7	2.9	1.2	0.1	97
321 - , meat and skin	0	0	0	0	3.5	5.8	2.3	0.1	99
322 - , weighed with bone	0	0	0	0	2.3	3.9	1.5	0.1	66
Grilled and fried									
323 **Breast**, *grilled without skin*, meat only	0	0	0	0	0.6	1.0	0.4	Tr	94
324 *grilled with skin*, meat only	0	0	0	0	0.9	1.3	0.6	Tr	84
325 - , meat and skin	0	0	0	0	0.9	1.4	0.7	Tr	90
326 **Breast**, strips, *stir-fried*	0	0	0	0	N	N	N	N	87
327 **Portions**, *deep-fried*, meat and skin	0	0	0	0	4.6	8.0	3.0	0.3	135
328 - , weighed with bone	0	0	0	0	3.2	5.5	2.0	0.2	94

Chicken continued

Inorganic constituents per 100g food

No. 18-	Food	Na	K	Ca	Mg	P	Fe	Cu	Zn	Cl	Mn	Se	I
						mg						µg	
	Casseroled continued												
317	**Thighs**, *casseroled*, meat and skin	70	170	26	16	170	1.0	0.07	1.8	70	0.02	15	7
318	-, -, weighed with bone	60	140	22	13	140	0.8	0.06	1.5	60	0.02	12	6
319	diced, *casseroled*, meat only	40	150	14	19	150	1.0	0.15	2.1	81	0.02	17	7
320	**Wing quarter**, *casseroled*, meat only	65	250	15	23	200	0.6	0.06	1.3	60	0.01	14	7
321	-, meat and skin	60	230	15	21	180	0.6	0.07	1.3	60	0.01	13	7
322	-, -, weighed with bone	40	160	10	14	120	0.4	0.05	0.9	40	0.01	9	5
	Grilled and fried												
323	**Breast**, *grilled without skin, meat only*	55	460	6	36	310	0.4	0.04	0.8	67	0.01	16	7
324	*grilled with skin*, meat only	60	410	7	34	280	0.5	0.06	0.8	66	0.01	15	6
325	-, meat and skin	61	400	7	33	270	0.6	0.06	0.8	69	0.01	15	6
326	**Breast**, strips, *stir-fried*	61	420	6	33	280	0.5	0.08	0.8	63	0.01	15	7
327	**Portions**, *deep-fried*, meat and skin	150	310	15	25	220	1.0	0.11	2.0	94	0.02	18	8
328	-, -, weighed with bone	110	210	10	17	150	0.7	0.07	1.3	65	0.01	13	6

Chicken continued

No. Food 18-	Retinol µg	Carotene µg	Vitamin D µg	Vitamin E mg	Thiamin mg	Ribo- flavin mg	Niacin mg	Trypt 60 mg	Vitamin B6 mg	Vitamin B12 µg	Folate µg	Panto- thenate mg	Biotin µg	Vitamin C mg
Casseroled continued														
317 **Thighs,** *casseroled,* meat and skin	30	Tr	0.2	0.34	0.05	0.11	4.9	3.9	0.20	Tr	8	1.04	3	0
318 -, -, weighed with bone	25	Tr	0.2	0.28	0.04	0.09	4.1	3.3	0.17	Tr	7	0.86	2	0
319 diced, *casseroled,* meat only	20	Tr	0.2	0.12	0.05	0.15	3.4	5.0	0.18	1	6	1.34	4	0
320 **Wing quarter,** *casseroled,* meat only	5	Tr	0.1	0.16	0.06	0.13	7.8	5.3	0.33	Tr	7	1.31	3	0
321 -, meat and skin	16	Tr	0.2	0.16	0.06	0.12	7.0	4.5	0.29	Tr	6	1.18	2	0
322 -, -, weighed with bone	11	Tr	0.1	0.11	0.04	0.08	4.7	3.0	0.19	Tr	4	0.79	2	0
Grilled and fried														
323 **Breast,** *grilled without skin,* meat only	Tr	Tr	0.3	0.17	0.14	0.13	15.8	6.2	0.63	Tr	6	1.67	2	0
324 *grilled with skin,* meat only	Tr	Tr	0.1	0.12	0.10	0.20	14.3	5.8	0.42	Tr	5	1.66	2	0
325 -, meat and skin	5	Tr	0.2	0.12	0.10	0.19	13.6	5.5	0.40	Tr	5	1.57	2	0
326 **Breast,** strips, *stir-fried*	Tr	Tr	0.2	N	0.11	0.16	14.4	5.8	0.44	Tr	5	1.56	2	0
327 **Portions,** *deep-fried,* meat and skin	28	Tr	0.3	0.32	0.06	0.22	7.5	4.9	0.34	1	14	1.34	3	0
328 -, -, weighed with bone	19	Tr	0.2	0.22	0.04	0.15	5.1	3.4	0.23	Tr	10	0.92	2	0

Chicken *continued*

Composition of food per 100g

No. 18-	Food	Description and main data sources	Edible Proportion	Water g	Total Nitrogen g	Protein g	Fat g	Carbo-hydrate g	Energy value kcal	kJ
	Roasted and barbecued									
329	**Dark meat**, *roasted*	19 samples of a mixture of fresh and frozen chicken	1.00	63.9	3.90	24.4	10.9	0	196	819
330	**Light meat**, *roasted*	19 samples of a mixture of fresh and frozen chicken	1.00	66.9	4.83	30.2	3.6	0	153	645
331	**Meat**, average, *roasted*	Calculated from 46% light meat and 54% dark meat from fresh and frozen chicken	1.00	65.3	4.37	27.3	7.5	0	177	742
332	**Skin**, dry, *roasted/grilled*	34 samples; crisply roasted	1.00	31.1	3.45	21.5	46.1	0	501	2070
333	-, moist, *roasted/grilled*	32 samples	1.00	41.6	2.72	17.0	42.6	0	452	1867
334	**Drumsticks**, *roasted*, meat only	20 samples of a mixture of fresh and frozen drumsticks	1.00	66.2	4.25	26.6	5.1	0	152	640
335	-, meat and skin	Calculated from 89% dark meat and 11% skin from fresh and frozen chicken	1.00	63.0	4.14	25.8	9.1	0	185	775
336	-, -, weighed with bone	Calculated from no. 335	0.63	39.7	2.60	16.2	5.7	0	116	486
337	**Leg quarter**, *roasted*, meat and skin	20 samples	1.00	60.9	3.34	20.9	16.9	0	236	981
338	-, -, weighed with bone	Calculated from no. 337	0.51	31.1	1.71	10.7	8.6	0	120	500
339	**Wing quarter**, *roasted*, meat and skin	20 samples	1.00	59.9	3.97	24.8	14.1	0	226	943
340	-, -, weighed with bone	Calculated from no. 339	0.53	31.7	2.10	13.1	7.5	0	120	500
341	**Whole chicken**, *roasted*, meat and skin	Calculated from 40% light meat, 47% dark meat and 13% skin from fresh and frozen chicken	1.00	61.3	4.21	26.3	12.5	0	218	910
342	-, -, weighed with bone	Calculated from no. 341	0.63	38.6	2.66	16.6	7.9	0	138	575

Chicken continued

Roasted and barbecued

No. 18-	Food	Starch g	Total sugars g	Dietary fibre Southgate method g	Dietary fibre Englyst method g	Fatty acids cis & trans Satd g	Mono-unsatd g	Poly-unsatd g	Total trans g	Cholesterol mg
329	**Dark meat**, *roasted*	0	0	0	0	2.9	5.1	2.2	0.1	120
330	**Light meat**, *roasted*	0	0	0	0	1.0	1.6	0.7	Tr	82
331	**Meat**, average, *roasted*	0	0	0	0	2.1	3.4	1.5	0.1	105
332	**Skin**, dry, *roasted/grilled*	0	0	0	0	12.9	22.5	7.7	0.6	170
333	-, moist, *roasted/grilled*	0	0	0	0	12.0	20.8	7.1	0.5	135
334	**Drumsticks**, *roasted*, meat only	0	0	0	0	1.4	2.4	1.0	Tr	135
335	-, meat and skin	0	0	0	0	2.5	4.3	1.8	0.1	135
336	-, -, weighed with bone	0	0	0	0	1.5	2.7	1.1	Tr	86
337	**Leg quarter**, *roasted*, meat and skin	0	0	0	0	4.6	7.8	3.2	0.1	115
338	-, -, weighed with bone	0	0	0	0	2.3	4.0	1.6	0.1	60
339	**Wing quarter**, *roasted*, meat and skin	0	0	0	0	3.9	6.4	2.7	0.1	100
340	-, -, weighed with bone	0	0	0	0	2.1	3.4	1.5	0.1	53
341	**Whole chicken**, *roasted*, meat and skin	0	0	0	0	3.4	5.7	2.4	0.1	110
342	-, -, weighed with bone	0	0	0	0	2.2	3.6	1.5	0.1	71

Inorganic constituents per 100g food

No. 18-	Food	Na	K	Ca	Mg	P	Fe	Cu	Zn	Cl	Mn	Se	I
						mg						µg	
	Roasted and barbecued												
329	**Dark meat**, *roasted*	100	300	17	23	200	0.8	0.08	2.2	88	0.02	17	6
330	**Light meat**, *roasted*	60	360	7	30	250	0.4	0.17	0.8	62	0.01	14	7
331	**Meat**, average, *roasted*	80	330	11	26	220	0.7	0.10	1.5	75	0.02	16	7
332	**Skin**, dry, *roasted/grilled*	80	260	16	26	210	1.3	0.05	1.2	N	0.03	N	N
333	-, moist, *roasted/grilled*	80	240	12	20	170	1.0	0.04	0.9	N	0.02	N	N
334	**Drumsticks**, *roasted*, meat only	170	260	16	25	210	0.9	0.10	2.5	84	0.02	17	7
335	-, meat and skin	130	280	15	25	210	1.0	0.09	2.3	90	0.02	17	7
336	-, -, weighed with bone	80	180	9	16	130	0.6	0.06	1.4	60	0.01	11	4
337	**Leg quarter**, *roasted*, meat and skin	95	230	12	20	180	0.8	0.06	1.7	85	0.01	16	7
338	-, -, weighed with bone	50	120	6	10	90	0.4	0.03	0.9	45	0.01	8	4
339	**Wing quarter**, *roasted*, meat and skin	100	260	11	24	200	0.6	0.04	1.1	75	0.01	15	7
340	-, -, weighed with bone	55	140	6	13	110	0.3	0.02	0.6	40	0.01	8	4
341	**Whole chicken**, *roasted*, meat and skin	80	320	11	26	220	0.7	0.09	1.5	80	0.02	15	7
342	-, -, weighed with bone	50	200	7	16	140	0.4	0.06	0.9	50	0.01	9	4

Chicken *continued*

Roasted and barbecued

No. 18-	Food	Retinol µg	Carotene µg	Vitamin D µg	Vitamin E mg	Thiamin mg	Ribo- flavin mg	Niacin mg	Trypt 60 mg	Vitamin B6 mg	Vitamin B12 µg	Folate µg	Panto- thenate mg	Biotin µg	Vitamin C mg
329	**Dark meat**, *roasted*	24	Tr	0.1	0.23	0.07	0.11	6.2	5.3	0.27	1	10	1.34	4	0
330	**Light meat**, *roasted*	Tr	Tr	0.3	0.31	0.07	0.23	12.6	5.5	0.54	Tr	10	1.38	2	0
331	**Meat**, average, *roasted*	11	Tr	0.2	0.23	0.07	0.16	9.2	5.3	0.36	Tr	10	1.39	3	0
332	**Skin**, dry, *roasted/grilled*	N	Tr	1.0	N	N	N	N	N	N	N	N	N	N	0
333	-, moist, *roasted/grilled*	N	Tr	0.6	N	N	N	N	N	N	N	N	N	N	0
334	**Drumsticks**, *roasted*, meat only	20	Tr	0.2	0.22	0.09	0.15	5.5	5.0	0.19	1	13	1.39	4	0
335	-, meat and skin	24	Tr	0.2	0.21	0.09	0.14	5.5	4.9	0.19	1	12	1.31	3	0
336	-, weighed with bone	15	Tr	0.1	0.13	0.06	0.09	3.5	3.1	0.12	Tr	8	0.82	2	0
337	**Leg quarter**, *roasted*, meat and skin	26	Tr	0.2	0.23	0.07	0.28	5.0	4.6	0.40	1	11	1.21	3	0
338	-, weighed with bone	13	Tr	0.1	0.12	0.04	0.14	2.5	2.4	0.20	Tr	6	0.62	2	0
339	**Wing quarter**, *roasted*, meat and skin	13	Tr	0.4	0.27	0.06	0.17	10.0	4.9	0.25	Tr	6	0.67	1	0
340	-, weighed with bone	7	Tr	0.2	0.14	0.03	0.09	5.3	2.6	0.13	Tr	3	0.35	1	0
341	**Whole chicken**, *roasted*, meat and skin	18	Tr	0.3	0.21	0.07	0.15	8.7	4.9	0.34	Tr	9	1.29	3	0
342	-, weighed with bone	11	Tr	0.2	0.13	0.04	0.09	5.5	3.1	0.21	Tr	6	0.81	2	0

Composition of food per 100g

No. 18-	Food	Description and main data sources	Edible Proportion	Water g	Total Nitrogen g	Protein g	Fat g	Carbo-hydrate g	Energy value kcal	kJ
	Roasted and barbecued *continued*									
343	**Whole chicken corn-fed,** *roasted,* dark meat only	6 samples	1.00	66.1	3.84	24.0	9.8	0	185	772
344	-, light meat only	6 samples	1.00	67.4	4.13	25.8	4.2	0	141	595
345	-, meat only	Calculated from 48% light meat and 52% dark meat	1.00	66.7	3.98	24.9	7.2	0	164	690
346	-, meat and skin	Calculated from 40% light meat, 43% dark meat and 17% skin	1.00	61.7	3.84	24.0	13.2	0	215	896
347	-, -, weighed with bone	Calculated from no. 346	0.60	37.0	2.30	14.4	7.9	0	129	537

No. Food	Starch	Total sugars	Dietary fibre Southgate method	Englyst method	Fatty acids cis & trans Satd	Mono-unsatd	Poly-unsatd	Total trans	Cholesterol
18-	g	g	g	g	g	g	g	g	mg
Roasted and barbecued continued									
Whole chicken corn-fed,									
roasted, dark meat only									
343	0	0	0	0	2.8	4.3	2.0	Tr	120
344 -, light meat only	0	0	0	0	1.2	1.8	0.9	Tr	76
345 -, meat only	0	0	0	0	2.1	3.1	1.5	Tr	99
346 -, meat and skin	0	0	0	0	3.8	5.8	2.6	0.1	110
347 -, -, weighed with bone	0	0	0	0	2.7	4.1	1.8	0.1	65

Inorganic constituents per 100g food

No. Food 18-	Na	K	Ca	Mg	P	Fe	Cu	Zn	Cl	Mn	Se	I
						mg					μg	
Roasted and barbecued continued												
343 **Whole chicken corn-fed,** *roasted*, dark meat only	95	290	9	22	200	1.0	0.08	2.1	76	0.02	16	6
344 -, light meat only	90	330	9	28	230	0.4	0.06	0.8	54	0.01	13	6
345 -, meat only	90	310	9	25	210	0.7	0.07	1.5	65	0.02	15	6
346 -, meat and skin	90	300	10	25	210	0.8	0.07	1.4	75	0.02	15	7
347 -, -, weighed with bone	55	180	6	15	130	0.5	0.04	0.8	45	0.01	9	4

No. Food 18-	Retinol µg	Carotene µg	Vitamin D µg	Vitamin E mg	Thiamin mg	Ribo-flavin mg	Niacin mg	Trypt 60 mg	Vitamin B6 mg	Vitamin B12 µg	Folate µg	Panto-thenate mg	Biotin µg	Vitamin C mg
Roasted and barbecued continued														
343 Whole chicken corn-fed, *roasted,* dark meat only	16	Tr	0.2	0.20	0.07	0.25	6.8	4.7	0.29	Tr	6	1.26	3	0
344 -, light meat only	Tr	Tr	0.2	0.14	0.06	0.13	12.0	5.1	0.54	Tr	10	1.35	2	0
345 -, meat only	8	Tr	0.2	0.17	0.07	0.19	9.3	4.9	0.41	Tr	8	1.30	3	0
346 -, meat and skin	16	Tr	0.3	0.17	0.07	0.18	8.7	4.4	0.37	Tr	7	1.20	2	0
347 -, -, weighed with bone	10	Tr	0.2	0.10	0.04	0.11	5.2	2.6	0.22	Tr	4	0.72	1	0

Turkey

Composition of food per 100g

No. Food 18-	Description and main data sources	Edible Proportion	Water g	Total Nitrogen g	Protein g	Fat g	Carbohydrate g	Energy value kcal	kJ
Raw									
348 **Dark meat**, *raw*	20 samples	1.00	75.8	3.26	20.4	2.5	0	104	439
349 **Light meat**, *raw*	20 samples	1.00	74.9	3.90	24.4	0.8	0	105	444
350 **Meat**, average, *raw*	Calculated from 56% light meat and 44% dark meat	1.00	75.3	3.62	22.6	1.6	0	105	443
351 **Skin**, *raw*	21 samples	1.00	53.4	2.24	14.0	30.7	0	332	1374
352 **Whole turkey**, *raw*	Calculated from 49% light meat, 39% dark meat and 13% skin	1.00	72.6	3.46	21.6	5.2	0	133	560
353 -, -, weighed with bone	Calculated from no. 352	0.70	50.8	2.42	15.1	3.6	0	93	390
Casseroled and stewed									
354 **Mince**, *stewed*	5 samples	1.00	65.3	4.58	28.6	6.8	0	176	739
355 **Thigh**, diced, *casseroled*, meat only	8 samples; skinless	1.00	64.5	4.54	28.3	7.5	0	181	760
Grilled and stir-fried									
356 **Breast**, fillet, *grilled*, meat only	9 samples; skinless	1.00	63.0	5.60	35.0	1.7	0	155	658
357 **strips**, *stir-fried*	8 samples; skinless	1.00	64.4	4.96	31.0	4.5	0	164	692

No. Food 18-	Starch g	Total sugars g	Dietary fibre Southgate method g	Dietary fibre Englyst method g	Fatty acids Satd g	cis & trans Mono-unsatd g	cis & trans Poly-unsatd g	Total trans g	Cholesterol mg
Raw									
348 **Dark meat**, *raw*	0	0	0	0	0.8	1.0	0.6	Tr	86
349 **Light meat**, *raw*	0	0	0	0	0.3	0.3	0.2	Tr	57
350 **Meat**, average, *raw*	0	0	0	0	0.5	0.6	0.4	Tr	70
351 **Skin**, *raw*	0	0	0	0	10.1	11.9	6.7	0.4	135
352 **Whole turkey**, *raw*	0	0	0	0	1.7	1.9	1.2	Tr	78
353 -, -, weighed with bone	0	0	0	0	1.1	1.3	0.8	Tr	55
Casseroled and stewed									
354 **Mince**, *stewed*	0	0	0	0	2.0	2.2	2.1	0.1	120
355 **Thigh**, diced, *casseroled*, meat only	0	0	0	0	2.5	2.7	1.8	0.1	120
Grilled and stir-fried									
356 **Breast**, fillet, *grilled*, meat only	0	0	0	0	0.6	0.6	0.3	Tr	74
357 **strips**, *stir-fried*	0	0	0	0	N	N	N	N	72

No. Food 18-	Na	K	Ca	Mg	P	Fe	Cu	Zn	Cl	Mn	Se	I
						mg					µg	
Raw												
348 **Dark meat**, *raw*	90	310	7	22	200	1.0	0.04	3.1	73	Tr	15	5
349 **Light meat**, *raw*	50	360	4	27	230	0.3	0.01	1.0	39	Tr	10	6
350 **Meat**, average, *raw*	68	340	5	25	220	0.6	0.02	1.9	54	Tr	13	6
351 **Skin**, *raw*	67	180	11	13	130	0.7	0.09	1.1	62	0.02	8	8
352 **Whole turkey**, *raw*	70	320	6	23	210	0.6	0.03	1.8	55	Tr	12	6
353 -, -, weighed with bone	50	230	4	16	150	0.4	0.02	1.3	40	Tr	8	4
Casseroled and stewed												
354 **Mince**, *stewed*	45	210	15	21	180	1.4	0.19	3.8	85	0.02	19	8
355 **Thigh**, diced, *casseroled*, meat only	65	230	12	23	210	2.0	0.20	5.4	84	0.02	19	8
Grilled and stir-fried												
356 **Breast**, fillet, *grilled*, meat only	90	550	5	42	380	0.6	0.08	1.7	85	0.01	17	8
357 **strips**, *stir-fried*	60	420	5	32	280	0.4	0.04	1.3	75	0.01	15	7

No. Food 18-	Retinol µg	Carotene µg	Vitamin D µg	Vitamin E mg	Thiamin mg	Ribo-flavin mg	Niacin mg	Trypt 60 mg	Vitamin B6 mg	Vitamin B12 µg	Folate µg	Panto-thenate mg	Biotin µg	Vitamin C mg
Raw														
348 **Dark meat**, *raw*	Tr	Tr	0.4	Tr	0.08	0.31	4.6	4.0	0.35	2	28	0.75	2	0
349 **Light meat**, *raw*	Tr	Tr	0.3	Tr	0.06	0.15	10.7	4.3	0.81	1	9	0.66	1	0
350 **Meat**, average, *raw*	Tr	Tr	0.3	0.01	0.07	0.22	8.0	4.4	0.61	2	17	0.70	2	0
351 **Skin**, *raw*	53	Tr	0.8	Tr	0.05	0.07	3.9	N	0.15	N	3	N	N	0
352 **Whole turkey**, *raw*	7	Tr	0.4	0.01	0.07	0.20	7.5	4.0	0.55	1	16	0.69	2	0
353 -, -, weighed with bone	5	Tr	0.3	Tr	0.05	0.14	5.2	2.8	0.38	1	11	0.48	1	0
Casseroled and stewed														
354 **Mince**, *stewed*	Tr	Tr	0.5	Tr	0.05	0.20	4.7	5.6	0.32	2	4	1.05	3	0
355 **Thigh**, diced, *casseroled*, meat only	Tr	Tr	0.5	Tr	0.07	0.22	6.0	5.5	0.41	2	21	1.04	3	0
Grilled and stir-fried														
356 **Breast**, fillet, *grilled*, meat only	Tr	Tr	0.4	0.02	0.07	0.15	14.0	6.8	0.63	1	7	0.95	2	0
357 **strips**, *stir-fried*	Tr	Tr	0.3	N	0.07	0.12	13.5	6.1	0.69	1	8	0.84	2	0

Roasted

No. 18-	Food	Description and main data sources	Edible Proportion	Water g	Total Nitrogen g	Protein g	Fat g	Carbohydrate g	Energy value kcal	kJ
358	**Dark meat**, *roasted*	27 samples including self-basting turkey	1.00	64.3	4.71	29.4	6.6	0	177	745
359	**Light meat**, *roasted*	18 samples	1.00	65.1	5.39	33.7	2.0	0	153	648
360	-, self-basting, *roasted*	9 samples	1.00	64.2	5.07	31.7	4.0	0	163	688
361	**Meat**, *average, roasted*	Calculated from 51% light meat and 49% dark meat from fresh, frozen and self-basting turkey	1.00	64.6	4.99	31.2	4.6	0	166	701
362	**Skin**, *dry, roasted*	10 samples	1.00	29.5	3.06	29.9	40.2	0	481	1995
363	moist, *roasted*	18 samples	1.00	45.7	3.46	21.6	31.7	0	372	1542
364	**Drumsticks**, *roasted*, meat only	9 samples	1.00	65.0	4.40	27.5	5.8	0	162	683
365	-, meat and skin	Calculated from 92% dark meat and 8% skin	1.00	62.6	4.38	27.4	8.5	0	186	780
366	-, -, weighed with bone	Calculated from no. 365	0.72	45.1	3.15	19.7	6.1	0	134	561
367	**Whole turkey**, *roasted*	Calculated from 47% light meat, 45% dark meat and 8% skin from fresh, frozen and self-basting turkey	1.00	62.1	4.94	30.9	7.4	0	190	799
368	-, -, weighed with bone	Calculated from no. 367	0.65	40.4	3.22	20.1	4.8	0	124	519

Turkey *continued*

No. Food 18-	Starch g	Total sugars g	Dietary fibre Southgate method g	Englyst method g	Satd g	Mono-unsatd g	Poly-unsatd g	Total trans g	Cholesterol mg
Roasted									
358 **Dark meat**, *roasted*	0	0	0	0	2.0	2.4	1.7	0.1	120
359 **Light meat**, *roasted*	0	0	0	0	0.7	0.7	0.5	Tr	82
360 -, self-basting, *roasted*	0	0	0	0	1.2	1.5	1.0	0.1	78
361 **Meat**, average, *roasted*	0	0	0	0	1.4	1.7	1.1	0.1	100
362 **Skin**, dry, *roasted*	0	0	0	0	13.2	15.6	8.8	0.6	290
363 moist, *roasted*	0	0	0	0	10.5	12.3	7.0	0.5	210
364 **Drumsticks**, *roasted*, meat only	0	0	0	0	1.7	2.2	1.4	0.1	115
365 -, meat and skin	0	0	0	0	2.6	3.3	2.1	0.1	130
366 -, -, weighed with bone	0	0	0	0	1.8	2.3	1.5	0.1	92
367 **Whole turkey**, *roasted*	0	0	0	0	2.3	2.7	1.8	0.1	115
368 -, -, weighed with bone	0	0	0	0	1.5	1.8	1.1	0.1	74

Turkey *continued*

Inorganic constituents per 100g food

Roasted

No. 18-	Food	mg										μg	
		Na	K	Ca	Mg	P	Fe	Cu	Zn	Cl	Mn	Se	I
358	**Dark meat**, *roasted*	110	330	17	25	260	1.2	0.11	3.4	86	0.02	17	8
359	**Light meat**, *roasted*	50	400	6	30	260	0.5	0.05	1.4	52	0.01	14	8
360	-, self-basting, *roasted*	90	340	7	27	240	0.4	0.06	1.3	110	0.01	16	6
361	**Meat**, average, *roasted*	90	350	11	27	260	0.8	0.09	2.5	85	0.01	17	8
362	**Skin**, dry, *roasted*	110	330	20	33	250	1.6	0.07	1.8	N	0.03	N	N
363	moist, *roasted*	120	290	13	27	220	1.3	0.09	1.3	N	0.03	N	N
364	**Drumsticks**, *roasted*, meat only	110	310	15	24	230	1.4	0.13	3.9	82	0.02	18	7
365	-, meat and skin	110	310	15	25	230	1.4	0.13	3.7	90	0.02	18	8
366	-, -, weighed with bone	75	220	11	18	170	1.0	0.09	2.7	65	0.01	13	6
367	**Whole turkey**, *roasted*	90	350	12	27	260	0.9	0.09	2.4	90	0.02	17	8
368	-, -, weighed with bone	60	230	8	17	170	0.6	0.06	1.6	60	0.01	11	5

Turkey continued

Roasted

No. Food 18-	Retinol µg	Carotene µg	Vitamin D µg	Vitamin E mg	Thiamin mg	Ribo-flavin mg	Niacin mg	Trypt 60 mg	Vitamin B_6 mg	Vitamin B_{12} µg	Folate µg	Panto-thenate mg	Biotin µg	Vitamin C mg
358 **Dark meat**, *roasted*	Tr	Tr	0.3	Tr	0.05	0.25	7.2	5.7	0.44	2	20	1.06	3	0
359 **Light meat**, *roasted*	Tr	Tr	0.1	0.02	0.05	0.16	12.9	6.8	0.47	1	18	0.97	2	0
360 -, self-basting, *roasted*	Tr	Tr	0.4	0.34	0.06	0.14	13.9	6.4	0.57	1	11	0.83	2	0
361 **Meat**, average, *roasted*	Tr	Tr	0.3	0.06	0.06	0.19	10.3	6.2	0.49	1	17	0.98	2	0
362 **Skin**, dry, *roasted*	N	Tr	N	N	N	N	N	N	N	N	N	N	N	0
363 moist, *roasted*	N	Tr	N	N	N	N	N	N	N	N	N	N	N	0
364 **Drumsticks**, *roasted*, meat only	Tr	Tr	0.5	Tr	0.07	0.22	5.9	5.4	0.40	2	20	1.01	3	0
365 -, meat and skin	5	Tr	0.5	Tr	0.07	0.21	6.0	5.1	0.39	1	19	0.98	3	0
366 -, -, weighed with bone	Tr	Tr	0.4	Tr	0.05	0.15	4.3	3.7	0.28	1	14	0.71	2	0
367 **Whole turkey**, *roasted*	5	Tr	0.4	0.07	0.06	0.19	10.1	5.9	0.48	1	16	0.96	2	0
368 -, -, weighed with bone	Tr	Tr	0.3	0.04	0.04	0.12	6.6	3.8	0.31	1	10	0.62	1	0

Other poultry

Composition of food per 100g

No. 18-	Food	Description and main data sources	Edible Proportion	Water g	Total Nitrogen g	Protein g	Fat g	Carbo-hydrate g	Energy value kcal	kJ
369	**Duck**, *raw*, meat only	19 samples, meat from dressed carcase	1.00	74.8	3.15	19.7	6.5	0	137	575
370	-, weighed with fat, skin and bone	Calculated from no. 369	0.28	20.9	0.88	5.5	1.8	0	38	160
371	*raw*, meat, fat and skin	10 samples; meat, fat and skin = 0.57 of dressed carcase	1.00	49.7	2.10	13.1	37.3	0	388	1603
372	*roasted*, meat only	10 samples from dressed carcase	1.00	62.1	4.05	25.3	10.4	0	195	815
373	-, weighed with fat, skin and bone	Calculated from no. 372	0.21	13.0	0.85	5.3	2.2	0	41	172
374	*roasted*, meat, fat and skin	20 samples; meat, fat and skin = 0.42 of dressed carcase	1.00	42.6	3.20	20.0	38.1	0	423	1750
375	**Goose**, *raw*, meat, fat and skin	5 samples; meat, fat and skin = 0.66 of dressed carcase	1.00	50.2	2.64	16.5	32.8	0	361	1494
376	*roasted*, meat, fat and skin	5 samples; meat, fat and skin = 0.65 of dressed carcase	1.00	51.1	4.40	27.5	21.2	0	301	1252

No. 18-	Food	Starch g	Total sugars g	Dietary fibre Southgate method g	Dietary fibre Englyst method g	Fatty acids Satd g	cis & trans Mono-unsatd g	cis & trans Poly-unsatd g	Total trans g	Cholesterol mg
369	**Duck**, *raw*, meat only	0	0	0	0	2.0	3.2	1.0	0.1	110
370	-, weighed with fat, skin and bone	0	0	0	0	0.6	0.9	0.3	Tr	30
371	*raw*, meat, fat and skin	0	0	0	0	10.7	19.0	5.6	0.3	72
372	*roasted*, meat only	0	0	0	0	3.3	5.2	1.3	0.1	115
373	-, weighed with fat, skin and bone	0	0	0	0	0.7	1.1	0.3	Tr	25
374	*roasted*, meat, fat and skin	0	0	0	0	11.4	19.3	5.3	0.3	99
375	**Goose**, *raw*, meat, fat and skin	0	0	0	0	(9.5)	(17.3)	(3.7)	Tr	(80)
376	*roasted*, meat, fat and skin	0	0	0	0	(6.6)	(9.9)	(2.4)	Tr	(91)

Other poultry

Inorganic constituents per 100g food

No. 18-	Food	Na	K	Ca	Mg	P	Fe	Cu	Zn	Cl	Mn	Se	I
						mg						µg	
369	**Duck**, *raw*, meat only	110	290	12	19	200	2.4	0.34	1.9	98	Tr	N	N
370	-, weighed with fat, skin and bone	31	81	3	5	56	0.7	0.10	1.0	27	Tr	N	N
371	*raw*, meat, fat and skin	73	190	8	14	120	1.3	0.18	1.3	62	Tr	N	N
372	*roasted*, meat only	96	270	13	20	200	2.7	0.31	2.6	96	Tr	(22)	N
373	-, weighed with fat, skin and bone	20	57	3	4	42	0.6	0.07	0.6	20	Tr	(5)	N
374	*roasted*, meat, fat and skin	87	220	22	17	180	1.7	0.23	2.2	76	0.20	(22)	N
375	**Goose**, *raw*, meat, fat and skin	61	270	7	17	160	2.2	0.14	1.7	56	Tr	N	N
376	*roasted*, meat, fat and skin	80	320	10	23	220	3.3	0.15	2.6	80	0.01	N	N

No. 18-	Food	Retinol µg	Carotene µg	Vitamin D µg	Vitamin E mg	Thiamin mg	Ribo-flavin mg	Niacin mg	Trypt 60 mg	Vitamin B6 mg	Vitamin B12 µg	Folate µg	Panto-thenate mg	Biotin µg	Vitamin C mg
369	**Duck**, *raw*, meat only	(24)	Tr	N	0.02	0.36	0.45	5.3	4.2	0.34	3	25	1.60	6	0
370	-, weighed with fat, skin and bone	(7)	Tr	N	Tr	0.10	0.13	1.5	1.2	0.10	1	7	0.45	2	0
371	*raw*, meat, fat and skin	(51)	Tr	N	N	0.14	0.51	3.5	2.4	0.33	2	7	(0.95)	N	0
372	*roasted*, meat only	N	N	N	0.02	0.26	0.47	5.1	5.4	0.25	3	10	1.50	4	0
373	-, weighed with fat, skin and bone	N	N	N	Tr	0.06	0.10	1.1	1.1	0.05	1	2	0.32	1	0
374	*roasted*, meat, fat and skin	N	N	N	N	0.18	0.51	3.8	4.2	0.31	2	15	2.60	7	0
375	**Goose**, *raw*, meat, fat and skin	(17)	Tr	N	N	0.12	0.30	3.6	(2.3)	0.35	2	8	N	N	0
376	*roasted*, meat, fat and skin	(21)	Tr	N	N	0.12	0.51	4.6	(5.5)	0.42	2	12	1.40	3	0

No. 18-	Food	Description and main data sources	Edible Proportion	Water g	Total Nitrogen g	Protein g	Fat g	Carbohydrate g	Energy value kcal	kJ
377	**Grouse**, *roasted*, meat only	10 samples, meat from dressed carcase	1.00	69.0	4.42	27.6	2.0	0	128	543
378	-, -, weighed with bone	Calculated from no. 377	0.46	31.7	2.03	12.7	0.9	0	59	249
379	**Hare**, *stewed*, meat only	Meat from dressed carcase	1.00	60.7	4.78	29.9	8.0	0	192	804
380	-, -, weighed with bone	Calculated from no. 379	0.73	44.3	3.48	21.8	5.8	0	139	585
381	**Partridge**, *roasted*, meat only	Meat from dressed carcase	1.00	54.5	5.87	36.7	7.2	0	212	890
382	-, -, weighed with bone	Calculated from no. 381	0.60	32.7	3.52	22.0	4.3	0	127	533
383	**Pheasant**, *roasted*, meat only	10 samples from dressed carcase	1.00	59.4	4.46	27.9	12.0	0	220	918
384	-, -, weighed with bone	Calculated from no. 383	0.52	30.9	2.32	14.5	6.2	0	114	476
385	**Pigeon**, *roasted*, meat only	10 samples, meat from dressed carcase	1.00	62.2	4.64	29.0	7.9	0	187	785
386	-, -, weighed with bone	Calculated from no. 385	0.47	29.2	2.18	13.6	3.7	0	88	368
387	**Rabbit**, *raw*, meat only	10 samples from leg and loin	1.00	71.5	3.50	21.9	5.5	0	137	576
388	*stewed*, meat only	30 samples of a mixture of fresh, wild, farmed and frozen imported	1.00	70.7	3.39	21.2	3.2	0	114	479
389	-, -, weighed with bone	Calculated from no. 388	0.60	42.4	2.03	12.7	1.9	0	68	286
390	**Venison**, *raw*	Haunch, meat only	1.00	74.4	3.55	22.2	1.6	0	103	437
391	*roasted*	Calculated from no. 390	1.00	60.4	5.70	35.6	2.5	0	165	698

Game

No. 18-	Food	Starch g	Total sugars g	Dietary fibre Southgate method g	Englyst method g	Fatty acids cis & trans Satd g	Mono-unsatd g	Poly-unsatd g	Total trans g	Cholesterol mg
377	**Grouse**, *roasted*, meat only	0	0	0	0	0.5	0.2	1.2	Tr	N
378	-, -, weighed with bone	0	0	0	0	0.2	0.1	0.5	Tr	N
379	**Hare**, *stewed*, meat only	0	0	0	0	N	N	N	Tr	N
380	-, -, weighed with bone	0	0	0	0	N	N	N	Tr	N
381	**Partridge**, *roasted*, meat only	0	0	0	0	1.9	3.3	1.7	Tr	N
382	-, -, weighed with bone	0	0	0	0	1.1	2.0	1.0	Tr	N
383	**Pheasant**, *roasted*, meat only	0	0	0	0	4.1	5.6	1.6	0.1	(220)
384	-, -, weighed with bone	0	0	0	0	2.1	2.9	0.8	Tr	(115)
385	**Pigeon**, *roasted*, meat only	0	0	0	0	N	N	N	Tr	N
386	-, -, weighed with bone	0	0	0	0	N	N	N	Tr	N
387	**Rabbit**, *raw*, meat only	0	0	0	0	2.1	1.3	1.8	0.1	53
388	*stewed*, meat only	0	0	0	0	1.7	0.7	0.6	0.1	49
389	-, -, weighed with bone	0	0	0	0	1.0	0.4	0.4	Tr	49
390	**Venison**, *raw*	0	0	0	0	0.8	0.4	0.4	0.1	50
391	*roasted*	0	0	0	0	N	N	N	Tr	N

Game

Inorganic constituents per 100g food

No. 18-	Food	Na	K	Ca	Mg	P	Fe	Cu	Zn	Cl	Mn	Se (µg)	I (µg)
						mg							
377	**Grouse**, *roasted*, meat only	110	380	45	32	260	4.6	0.34	1.6	360	0.29	(20)	N
378	-, -, weighed with bone	51	180	21	15	120	2.1	0.16	0.8	170	0.13	(9)	N
379	**Hare**, *stewed*, meat only	40	210	21	22	250	10.8	N	N	74	N	N	N
380	-, -, weighed with bone	29	150	15	16	180	7.9	N	N	54	N	N	N
381	**Partridge**, *roasted*, meat only	100	410	46	36	310	7.7	N	N	99	N	N	N
382	-, -, weighed with bone	60	240	28	22	190	4.6	N	N	59	N	N	N
383	**Pheasant**, *roasted*, meat only	66	360	28	26	220	2.2	0.10	1.3	170	0.02	(14)	N
384	-, -, weighed with bone	34	190	15	13	110	1.1	0.05	0.7	88	0.01	N	N
385	**Pigeon**, *roasted*, meat only	92	400	32	30	310	7.2	0.33	1.7	87	0.05	N	N
386	-, -, weighed with bone	43	190	15	14	150	3.4	0.15	0.8	41	0.02	N	N
387	**Rabbit**, *raw*, meat only	67	360	22	25	220	1.0	0.06	1.4	74	0.01	17	N
388	*stewed*, meat only	48	200	39	18	150	1.1	0.06	1.7	45	0.02	(16)	N
389	-, -, weighed with bone	29	120	23	11	90	0.7	0.04	1.0	27	0.01	(10)	N
390	**Venison**, *raw*	55	340	5	25	210	3.3	0.21	2.4	64	0.03	9	N
391	*roasted*	52	290	6	27	240	5.1	0.36	3.9	59	0.04	N	N

Game

No. Food 18-	Retinol µg	Carotene µg	Vitamin D µg	Vitamin E mg	Thiamin mg	Ribo-flavin mg	Niacin mg	Trypt 60 mg	Vitamin B6 mg	Vitamin B12 µg	Folate µg	Panto-thenate mg	Biotin µg	Vitamin C mg
377 Grouse, *roasted*, meat only	N	N	N	N	0.19	0.80	7.0	5.8	0.64	1	37	N	N	0
378 -, -, weighed with bone	N	N	N	N	0.08	0.37	3.2	3.9	0.29	Tr	17	N	N	0
379 Hare, *stewed*, meat only	N	N	N	N	N	N	N	5.6	N	N	N	N	N	0
380 -, -, weighed with bone	N	N	N	N	N	N	N	4.1	N	N	N	N	N	0
381 Partridge, *roasted*, meat only	N	N	N	N	N	N	N	6.9	N	N	N	N	N	0
382 -, -, weighed with bone	N	N	N	N	N	N	N	4.1	N	N	N	N	N	0
383 Pheasant, *roasted*, meat only	N	N	N	N	0.02	0.29	9.2	6.0	0.57	3	20	(0.96)	N	0
384 -, -, weighed with bone	N	N	N	N	0.01	0.15	4.8	3.1	0.30	1	10	(0.50)	N	0
385 Pigeon, *roasted*, meat only	N	N	N	Tr	0.27	1.17	7.0	5.2	0.82	8	8	N	N	0
386 -, -, weighed with bone	N	N	N	Tr	0.13	0.55	3.3	2.3	0.38	4	4	N	N	0
387 Rabbit, *raw*, meat only	N	N	N	0.13	0.10	0.19	8.4	4.1	0.50	10	5	0.80	1	0
388 *stewed*, meat only	N	N	N	N	0.02	0.16	6.2	5.1	0.29	3	5	0.80	1	0
389 -, -, weighed with bone	N	N	N	N	0.01	0.10	6.2	2.7	0.17	2	3	0.48	1	0
390 Venison, *raw*	N	N	N	N	N	(0.25)	N	N	N	N	N	N	N	0
391 *roasted*	N	N	N	N	0.16	0.69	5.5	6.5	0.65	1	6	N	N	0

18-392 to 18-408

Composition of food per 100g

No. 18-	Food	Description and main data sources	Edible Proportion	Water g	Total Nitrogen g	Protein g	Fat g	Carbo-hydrate g	Energy value kcal	kJ
392	Brain, lamb, *boiled*	11 samples	1.00	77.0	1.86	11.6	8.8	0	126	523
393	Giblets, chicken, *raw*	Mixed sample including heart, liver and gizzard	(0.69)	72.9	2.74	17.1	8.2	0	142	594
394	*boiled*	Mixed sample including heart, liver and gizzard	0.69	67.7	3.66	22.9	7.1	0	156	652
395	turkey, *boiled*	Mixed sample including heart, liver, gizzard and neck	0.48	61.0	4.50	28.1	8.8	0	192	803
396	Heart, lamb, *raw*	12 samples	0.73	75.6	2.73	17.1	5.6	0	119	498
397	*roasted*	10 samples, fat and valves removed	1.00	58.8	4.05	25.3	13.9	0	226	944
398	ox, *raw*	10 samples	0.79	76.4	2.91	18.2	3.5	0	104	439
399	*stewed*	10 samples, fat and valves removed	1.00	64.3	4.45	27.8	5.1	0	157	661
400	pig, *raw*	12 samples	0.83	77.9	2.74	17.1	3.2	0	97	409
401	*stewed*	10 samples, fat and valves removed	1.00	66.3	4.02	25.1	6.8	0	162	678
402	Kidney, lamb, *raw*	10 samples	0.86	78.6	2.72	17.0	2.6	0	91	385
403	*fried*	10 samples, skin and core removed	1.00	62.8	3.79	23.7	10.3	0	188	784
404	ox, *raw*	12 samples	0.88	78.8	2.75	17.2	2.1	0	88	370
405	*stewed*	10 samples, skin and core removed	1.00	69.2	3.92	24.5	4.4	0	138	579
406	pig, *raw*	12 samples	0.90	80.0	2.48	15.5	2.7	0	86	363
407	*fried*	10 samples, skin and core removed	1.00	58.1	4.67	29.2	9.5	0	202	848
408	*stewed*	20 samples, core removed. Salt added	1.00	66.3	3.91	24.4	6.1	0	153	641

No. Food		Starch	Total sugars	Dietary fibre		Fatty acids cis & trans			Total trans	Cholesterol
				Southgate method	Englyst method	Satd	Mono-unsatd	Poly-unsatd		
18-		g	g	g	g	g	g	g	g	mg
392	**Brain, lamb**, *boiled*	0	0	0	0	2.0	1.7	1.0	Tr	2200
393	**Giblets, chicken**, *raw*	0	0	0	0	N	N	N	Tr	N
394	*boiled*	0	0	0	0	N	N	N	Tr	(390)
395	**turkey**, *boiled*	0	0	0	0	N	N	N	Tr	N
396	**Heart, lamb**, *raw*	0	0	0	0	2.1	1.7	0.5	Tr	140
397	*roasted*	0	0	0	0	N	N	N	Tr	260
398	**ox**, *raw*	0	0	0	0	1.7	0.9	0.1	Tr	140
399	*stewed*	0	0	0	0	2.5	1.3	0.1	Tr	230
400	**pig**, *raw*	0	0	0	0	1.2	0.9	0.4	Tr	79
401	*stewed*	0	0	0	0	N	N	N	Tr	N
402	**Kidney, lamb**, *raw*	0	0	0	0	0.9	0.6	0.5	Tr	315
403	*fried*	0	0	0	0	N	N	N	Tr	610
404	**ox**, *raw*	0	0	0	0	0.8	0.4	0.4	Tr	265
405	*stewed*	0	0	0	0	1.4	1.0	0.9	0.1	460
406	**pig**, *raw*	0	0	0	0	0.9	0.7	0.4	Tr	410
407	*fried*	0	0	0	0	N	N	N	Tr	N
408	*stewed*	0	0	0	0	2.0	1.6	0.9	Tr	700

Inorganic constituents per 100g food

No. 18-	Food	mg										µg	
		Na	K	Ca	Mg	P	Fe	Cu	Zn	Cl	Mn	Se	I
392	**Brain, lamb**, *boiled*	210	190	(11)	15	320	1.4	0.23	1.4	250	N	N	N
393	**Giblets, chicken**, *raw*	82	250	10	17	170	6.7	0.34	2.8	130	0.20	N	N
394	*boiled*	67	210	10	18	280	6.8	0.21	3.5	93	0.20	N	N
395	**turkey**, *boiled*	63	210	34	20	230	5.8	0.30	5.6	81	0.14	N	N
396	**Heart, lamb**, *raw*	140	280	7	21	210	3.6	0.52	2.0	140	0.02	(2)	N
397	*roasted*	84	210	7	21	240	6.0	0.66	2.8	100	0.03	N	N
398	**ox**, *raw*	88	290	5	22	210	5.0	0.37	1.8	44	0.04	(3)	N
399	*stewed*	65	200	9	25	250	6.2	0.78	3.3	83	0.06	(3)	N
400	**pig**, *raw*	80	300	6	20	220	4.8	0.37	1.8	110	0.02	(5)	N
401	*stewed*	76	190	10	20	220	5.3	0.60	2.9	63	0.03	N	N
402	**Kidney, lamb**, *raw*	150	260	8	18	270	5.5	0.62	2.5	230	0.16	150	N
403	*fried*	230	280	14	21	350	11.2	0.58	3.6	410	0.13	88	N
404	**ox**, *raw*	170	240	11	16	240	7.2	0.48	1.9	300	0.11	150	15
405	*stewed*	150	210	17	19	290	9.0	0.63	3.0	190	0.14	(210)	N
406	**pig**, *raw*	200	220	10	17	250	6.4	0.43	2.4	330	0.11	150	7
407	*fried*	220	400	12	28	430	9.1	1.08	4.4	430	0.26	(270)	N
408	*stewed*	370	190	13	21	330	6.4	0.84	4.7	480	0.18	(250)	N

Offal

No. 18-	Food	Retinol µg	Carotene µg	Vitamin D µg	Vitamin E mg	Thiamin mg	Ribo-flavin mg	Niacin mg	Trypt 60 mg	Vitamin B6 mg	Vitamin B12 µg	Folate µg	Panto-thenate mg	Biotin µg	Vitamin C mg
392	**Brain, lamb**, *boiled*	Tr	Tr	Tr	1.10	0.10	0.24	2.1	2.5	0.08	8	6	1.40	3	17
393	**Giblets, chicken**, *raw*	5500	Tr	N	0.71	0.11	0.13	5.9	N	0.29	14	530	1.10	N	12
394	*boiled*	N	N	N	N	0.16	2.15	5.2	N	0.48	15	420	2.70	130	6
395	**turkey**, *boiled*	3100	Tr	N	0.07	0.09	1.14	5.2	N	0.26	8	330	1.50	53	3
396	**Heart, lamb**, *raw*	Tr	Tr	N	0.37	0.48	0.90	6.9	3.6	0.29	8	2	2.50	4	7
397	*roasted*	Tr	Tr	N	N	0.24	1.37	3.8	5.6	0.26	6	2	(3.80)	(8)	2
398	**ox**, *raw*	Tr	Tr	N	0.45	0.45	0.80	6.3	4.0	0.23	13	4	2.40	2	7
399	*stewed*	Tr	Tr	N	0.72	0.21	1.20	4.9	(6.7)	0.05	13	5	1.70	5	2
400	**pig**, *raw*	Tr	Tr	N	(0.37)	(0.48)	(0.90)	(6.9)	3.7	(0.29)	(8)	(2)	(2.50)	(4)	5
401	*stewed*	Tr	Tr	N	N	0.15	1.14	4.7	N	0.31	2	3	N	N	2
402	**Kidney, lamb**, *raw*	96	Tr	N	0.37	0.34	2.38	9.2	3.5	0.59	17	8	4.30	37	9
403	*fried*	110	Tr	N	0.41	0.52	3.10	9.1	5.3	0.48	54	70	4.60	73	5
404	**ox**, *raw*	105	410	N	0.55	0.50	2.53	6.0	3.4	0.37	15	63	3.10	24	8
405	*stewed*	45	N	N	0.42	0.24	3.29	6.2	(5.5)	0.57	38	130	3.10	79	5
406	**pig**, *raw*	115	Tr	N	0.44	0.56	2.58	7.6	3.5	0.39	40	39	3.00	32	7
407	*fried*	N	Tr	N	N	0.41	3.70	14.9	5.3	0.22	28	30	4.30	129	12
408	*stewed*	46	Tr	N	0.36	0.19	2.10	6.1	5.2	0.28	15	43	2.40	53	11

No. 18-	Food	Description and main data sources	Edible Proportion	Water g	Total Nitrogen g	Protein g	Fat g	Carbohydrate g	Energy value kcal	kJ
409	**Liver, calf**, *raw*	10 samples	1.00	72.0	2.93	18.3	3.4	Tr	104	437
410	*fried*	10 samples	1.00	64.5	3.57	22.3	9.6	Tr	176	734
411	**chicken**, *raw*	10 samples	1.00	75.8	2.83	17.7	2.3	Tr	92	386
412	*fried*	10 samples	1.00	65.9	3.54	22.1	8.9	Tr	169	705
413	**lamb**, *raw*	10 samples	1.00	69.5	3.25	20.3	6.2	Tr	137	575
414	*fried*	10 samples	1.00	53.9	4.82	30.1	12.9	Tr	237	989
415	**ox**, *raw*	33 samples	1.00	68.6	3.37	21.1	7.8	Tr	155	647
416	*stewed*	18 samples, coated in seasoned flour	1.00	62.6	3.96	24.8	9.5	3.6	198	831
417	**pig**, *raw*	10 samples	1.00	70.5	3.41	21.3	3.1	Tr	113	477
418	*stewed*	18 samples, coated in seasoned flour	1.00	62.1	4.09	25.6	8.1	3.6	189	793
419	**Oxtail**, *raw*	12 samples	0.38	68.6	3.20	20.0	10.1	0	171	714
420	*stewed*	12 samples, meat only. Salt added	1.00	53.9	4.88	30.5	13.4	0	243	1014
421	**Sweetbread**, lamb, *raw*	Thymus and pancreas, 12 samples	1.00	75.5	2.44	15.3	7.8	0	131	549
422	*fried*	3 samples	1.00	57.7	4.59	28.7	11.4	0	217	910
423	**Tongue, lamb**, *raw*	20 samples	0.57	67.9	2.45	15.3	14.6	0	193	800
424	**ox**, *pickled, raw*	6 samples, fat and skin removed	0.60	62.4	2.51	15.7	17.5	0	220	914
425	*stewed*	Fat and skin removed	1.00	48.6	3.12	19.5	23.9	0	293	1216
426	**ox**, *stewed*	10 samples, fat and skin removed	1.00	58.9	3.10	19.4	18.3	0	242	1007
427	**sheep**, *stewed*	Fat and skin removed	1.00	56.9	2.91	18.2	24.0	0	289	1197
428	**Tripe**, *dressed, raw*	6 samples	1.00	92.1	1.14	7.1	0.5	0	33	139
429	**Trotters and tails**, *boiled*	23% trotters and 77% pig tails. Salt added	0.54	53.5	3.17	19.8	22.3	0	280	1162

No. 18-	Food	Starch g	Total sugars g	Dietary fibre Southgate method g	Englyst method g	Fatty acids cis & trans Satd g	Mono-unsatd g	Poly-unsatd g	Total trans g	Cholesterol mg
409	**Liver, calf,** *raw*	0	0	0	0	1.0	0.6	0.8	Tr	370
410	*fried*	0	0	0	0	N	N	N	Tr	330
411	**chicken,** *raw*	0	0	0	0	0.7	0.5	0.4	Tr	380
412	*fried*	0	0	0	0	N	N	N	Tr	350
413	**lamb,** *raw*	0	0	0	0	1.7	1.8	0.9	Tr	430
414	*fried*	0	0	0	0	N	N	N	Tr	400
415	**ox,** *raw*	0	0	0	0	2.9	1.3	1.6	Tr	270
416	*stewed*	3.6	Tr	0	0	3.5	1.5	2.0	Tr	240
417	**pig,** *raw*	0	0	0	0	1.0	0.5	0.9	Tr	260
418	*stewed*	3.6	0	0	0	2.5	1.3	2.2	Tr	290
419	**Oxtail,** *raw*	0	0	0	0	4.2	4.8	0.4	Tr	75
420	*stewed*	0	0	0	0	N	N	N	Tr	110
421	**Sweetbread,** lamb, *raw*	0	0	0	0	3.0	2.4	0.3	Tr	260
422	*fried*	0	0	0	0	N	N	N	Tr	N
423	**Tongue, lamb,** *raw*	0	0	0	0	N	N	N	Tr	180
424	**ox,** *pickled, raw*	0	0	0	0	N	N	N	Tr	78
425	*stewed*	0	0	0	0	N	N	N	Tr	(100)
426	**ox,** *stewed*	0	0	0	0	N	N	N	Tr	N
427	**sheep,** *stewed*	0	0	0	0	N	N	N	Tr	(270)
428	**Tripe,** *dressed, raw*	0	0	0	0	0.2	0.2	Tr	Tr	64
429	**Trotters and tails,** *boiled*	0	0	0	0	N	N	N	Tr	N

No. 18-	Food	Na	K	Ca	Mg	P	Fe (mg)	Cu	Zn	Cl	Mn	Se (µg)	I
409	**Liver, calf**, *raw*	71	310	6	20	320	11.5	20.48	14.2	85	0.24	(22)	N
410	*fried*	70	350	8	24	380	12.2	23.86	15.9	110	0.29	(27)	N
411	**chicken**, *raw*	76	260	8	19	280	9.2	0.50	3.7	130	0.31	N	N
412	*fried*	79	300	9	23	350	11.3	0.52	3.8	110	0.35	N	N
413	**lamb**, *raw*	73	290	6	19	390	7.5	9.67	4.0	120	0.32	42	5
414	*fried*	82	340	8	25	500	10.9	13.54	5.9	140	0.45	(62)	N
415	**ox**, *raw*	81	320	6	19	360	7.0	2.50	4.0	90	0.37	42	13
416	*stewed*	110	250	11	19	380	7.8	2.30	4.3	120	0.44	(50)	N
417	**pig**, *raw*	82	300	5	20	390	13.9	1.13	7.0	130	0.29	42	N
418	*stewed*	130	250	11	22	390	17.0	2.50	8.2	150	0.40	(50)	N
419	**Oxtail**, *raw*	110	270	9	20	160	2.7	0.20	5.6	110	N	N	N
420	*stewed*	190	170	14	18	140	3.8	0.27	8.8	270	N	N	N
421	**Sweetbread**, lamb, *raw*	75	420	8	21	400	1.7	0.20	1.9	120	0.04	N	N
422	*fried*	52	400	27	34	480	1.8	0.40	3.6	52	0.08	N	N
423	**Tongue, lamb**, *raw*	420	250	6	33	170	2.2	0.64	2.7	550	(0.05)	N	N
424	**ox**, *pickled, raw*	1210	300	7	19	150	4.9	0.37	3.5	1750	0.01	N	N
425	*stewed*	1000	150	31	16	230	3.0	N	N	1450	0.01	N	N
426	**ox**, *stewed*	860	140	9	13	140	2.5	0.09	3.8	1420	0.01	N	N
427	**sheep**, *stewed*	80	110	11	13	200	3.4	N	N	80	N	N	N
428	**Tripe**, *dressed, raw*	50	12	52	3	16	0.2	0.04	0.7	8	0.02	N	N
429	**Trotters and tails**, *boiled*	1620	30	130	8	110	0.7	0.07	2.4	2490	0.01	N	N

Offal *continued*

No. 18-	Food	Retinol μg	Carotene μg	Vitamin D μg	Vitamin E mg	Thiamin mg	Ribo-flavin mg	Niacin mg	Trypt 60 mg	Vitamin B6 mg	Vitamin B12 μg	Folate μg	Panto-thenate mg	Biotin μg	Vitamin C mg
409	**Liver, calf**, *raw*	18800[a]	100	0.3	0.45	0.22	2.52	12.5	4.3	0.48	68	155	8.40	39	21
410	*fried*	(25200)[a]	100	0.3	0.50	0.61	2.89	13.6	5.8	0.89	58	110	4.10	50	19
411	**chicken**, *raw*	9700[a]	Tr	0.2	0.60	0.48	2.16	10.6	4.1	0.82	35	995	6.10	210	28
412	*fried*	(10500)[a]	Tr	N	0.34	0.63	2.72	12.9	4.4	0.55	45	1350	5.90	216	23
413	**lamb**, *raw*	17300[a]	85	0.5	0.77	0.39	4.64	16.4	4.3	0.47	54	205	8.20	41	19
414	*fried*	(19700)[a]	60	0.5	0.32	0.38	5.65	19.9	4.9	0.53	83	260	8.00	33	19
415	**ox**, *raw*	14200[a]	1540	1.1	0.42	0.23	3.10	13.4	4.5	0.83	110	330	8.10	33	23
416	*stewed*	(17300)[a]	1540	1.1	0.44	0.18	3.60	10.3	5.3	0.52	110	290	5.70	50	15
417	**pig**, *raw*	17400[a]	Tr	1.1	0.33	0.42	3.36	14.8	4.6	0.59	23	295	6.50	27	21
418	*stewed*	(22600)[a]	Tr	1.1	0.16	0.21	3.10	11.5	5.5	0.64	26	110	4.60	34	9
419	**Oxtail**, *raw*	Tr	Tr	Tr	0.29	0.03	0.29	4.5	4.3	0.27	3	7	1.00	1	0
420	*stewed*	Tr	Tr	Tr	0.45	0.02	0.28	3.3	6.5	0.14	2	9	0.90	2	0
421	**Sweetbread**, lamb, *raw*	Tr	Tr	Tr	0.44	0.03	0.25	3.7	3.3	0.03	6	13	1.00	3	18
422	*fried*	Tr	Tr	Tr	N	0.04	0.29	2.3	4.1	0.04	3	9	0.80	4	23
423	**Tongue, lamb**, *raw*	Tr	Tr	Tr	0.21	0.17	0.49	4.9	3.3	0.17	7	4	1.00	1	7
424	**ox**, *pickled, raw*	Tr	Tr	Tr	0.28	0.10	0.38	6.4	3.4	0.18	5	6	0.80	2	3
425	*stewed*	Tr	Tr	Tr	(0.35)	(0.06)	(0.29)	(4.1)	4.2	(0.09)	(4)	(5)	(0.50)	(3)	(2)
426	**ox**, *stewed*	Tr	Tr	N	N	0.04	0.28	2.6	3.4	0.09	4	11	0.60	2	3
427	**sheep**, *stewed*	Tr	Tr	Tr	(0.32)	(0.13)	(0.45)	(3.7)	3.9	(0.10)	(7)	(4)	(0.80)	(2)	(6)
428	**Tripe**, *dressed, raw*	Tr	Tr	Tr	0.08	Tr	Tr	Tr	1.2	Tr	Tr	7	Tr	1	3
429	**Trotters and tails**, *boiled*	Tr	Tr	Tr	N	0.06	0.20	0.9	3.7	N	1	3	N	N	0

[a] Total retinol

117

Appendices

ALTERNATIVE AND TAXONOMIC NAMES

The main parts of the beef, lamb and pig carcase are shown in the diagrams. These are usually subdivided into the cuts given in this book, but because some items may be derived from more than one part of the animal (e.g. stewing beef or mince), and because many cuts are known by different names in different parts of the country, some of the more common alternative names and sources are given in the tables below. The main alternative names are also included in the index. The abbreviation 'spp' is used to indicate that one or more than one species of the specified Genus is included.

Food names	Taxonomic names
Beef	Bos taurus
Chicken	Gallus domesticus
Duck	Anas platyrhynchos
Goose	Anser anser
Grouse	Lagopus scroticus
Hare	Lepus europaeus
Lamb	Ovis aries
Partridge	Perdix perdix
Pheasant	Phasianus colchicus
Pigeon	Columba spp
Pork	Sus scrofa
Rabbit	Lepus cuniculus
Turkey	Meleagris gallopavo
Veal	Bos taurus
Venison	Cervus spp

BEEF

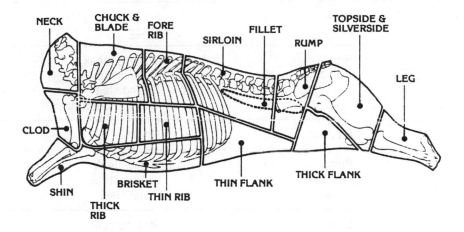

Beef	Alternative names and retail cuts
Brisket	Brisket joint Pot roast Slow roast
Chuck and blade	Braising steak Casserole steak
Clod	Diced beef Mince Stewing beef Stewing steak
Fillet	Fillet steak
Flank, thick	Beef olives Frying steak Mini joints Steak bars Top rump
Flank, thin	Mince Skirt steaks
Fore rib	Fore rib roast Rib eye steak
Leg	Stewing beef Stewing steak

Beef	Alternative names and retail cuts
Neck	Diced beef Mince
Rib, thick	Braising steak Frying steak Grilling steak 'Leg of mutton cut'
Rib, thin	Mince
Rump	Frying steak Grilling steak Rump steak
Shin	Stewing beef Stewing steak
Silverside	Silverside joint Lean cubes
Sirloin	Sirloin joint Sirloin steaks Striploin T-bone steak Wing rib
Topside	Roasting beef Topside joint Topside steaks

LAMB

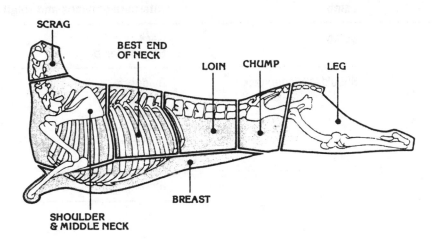

Lamb	Alternative names and retail cuts
Best end neck	Crown Cutlets Rack of lamb
Breast	Flank
Chump	Chump chops Chump end Chump steaks Gigot chops Saddle
Leg	Fillet end Gigot Knuckle half end Leg joint Leg mini joints Shank half end Stir fry
Loin	Barnsley chops Best loin chops Lamb chops Loin chops Medallions Noisettes Valentine steaks

Lamb *continued*

Lamb	Alternative names and retail cuts
Scrag	Neck Stewing lamb
Shoulder/Middle neck	Bladeside Knuckle end Mince Neck Neck fillet Shoulder joint Shoulder mini joints Stewing lamb

PORK

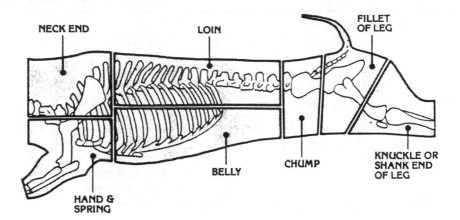

Pork	Alternative names and retail cuts
Belly	Belly joint Belly slices Flank end Streaky pork
Chump	Chump chops Chump end Chump steaks Gigot chops Pork steaks
Hand and spring	Fore leg roast Knuckle Mince Shank Shoulder
Leg	Diced pork Escalopes Fillet end Gigot Kebabs Knuckle end Leg joint Leg mini joints Leg steaks Nuggets Shank end

Pork *continued*

Pork	Alternative names and retail cuts
BLoin	Fillet Loin chops Loin joint Medallions Pork chops Single loin chops Valentine steaks
Neck end	Collar Crop of pork Spare rib chops Spare rib joints

PROPORTIONS OF TRIMMABLE LEAN, FAT AND INEDIBLE MATERIAL IN BEEF, LAMB AND PORK

Every sample of raw and cooked beef, veal, lamb and pork was separated before analysis into its lean, fat and inedible material (mainly bone, but also any gristle or crackling). The amounts were weighed, and the proportions are given below. It is important to realise that each set of values was derived from different samples. Small differences may therefore reflect differences between the samples as much as the effect of cooking.

	Number of samples	Means and ranges (%)		
		Trimmable lean	Trimmable fat	Inedible
BEEF				
Braising steak, *raw*	10	92 (82-100)	7 (0-18)	1 (0-6)
braised	10	90 (76-97)	9 (2-22)	0 (0-3)
slow-cooked	10	88 (82-100)	10 (0-15)	0 (0)
Brisket, *raw*	10	79 (50-92)	19 (6-49)	1 (0-11)
boiled	10	82 (45-96)	15 (2-52)	2 (0-17)
Fillet steak, *raw*	17	94 (89-98)	6 (0-11)	0 (0-2)
fried	16	96 (88-100)	4 (0-12)	0 (0-1)
grilled	20	97 (86-100)	3 (0-13)	0 (0-2)
from steakhouse	7	97 (94-100)	3 (0-6)	0 (0)
Flank, boneless, *raw*	6	76 (55-91)	23 (8-44)	0 (0-1)
pot-roasted	6	76 (59-92)	21 (8-39)	2 (0-8)
Fore-rib, *raw*	10	63 (39-81)	21 (10-35)	15 (2-31)
microwaved	10	64 (46-75)	18 (13-30)	17 (10-25)
roasted	10	65 (51-77)	18 (11-27)	16 (6-23)
Rump steak, *raw*	10	88 (76-98)	11 (1-23)	1 (0-11)
barbecued	10	91 (79-100)	8 (0-21)	0 (0)
fried	12	87 (79-100)	12 (0-19)	0 (0-3)
grilled	10	91 (75-99)	8 (1-25)	0 (0-1)
stir-fried	10	86 (72-100)	13 (0-27)	0 (0)
cooked from steakhouse	6	85 (81-90)	14 (9-18)	0 (0)
Silverside, *raw*	10	80 (72-92)	18 (7-28)	1 (0-5)
pot-roasted	10	85 (79-91)	14 (9-20)	0 (0)
salted, raw	8	80 (73-85)	19 (14-26)	0 (0-1)
boiled	8	88 (79-93)	11 (7-20)	0 (0)
Sirloin joint, *roasted*	10	82 (65-94)	16 (5-31)	0 (0-3)

	Number of samples	Means and ranges (%)		
		Trimmable lean	Trimmable fat	Inedible
Beef *continued*				
Sirloin steak, *raw*	16	80 (65-91)	18 (6-34)	1 (0-9)
fried	10	82 (75-91)	16 (7-25)	1 (0-9)
grilled, rare	10	83 (76-91)	16 (9-24)	0 (0-2)
-, medium-rare	19	86 (71-93)	12 (7-29)	1 (0-15)
-, well-done	10	86 (70-94)	13 (6-29)	1 (0-5)
from steakhouse	7	85 (64-94)	14 (6-35)	0 (0)
Stewing steak, *raw*	10	90 (83-98)	9 (1-16)	0 (0-3)
pressure-cooked	10	89 (74-100)	10 (0-25)	0 (0)
stewed	10	84 (67-100)	14 (0-31)	0 (0)
frozen, stewed	10	85 (71-91)	13 (7-27)	1 (0-7)
Topside, raw	10	84 (78-94)	15 (6-22)	1 (0-6)
microwaved [a]	10	89 (81-94)	11 (5-19)	0 (0)
roasted, medium-rare[a]	10	87 (76-92)	12 (7-23)	0 (0)
-, well-done[a]	10	88 (78-93)	11 (6-21)	0 (0)
VEAL				
Escalope, *raw*	9	99 (96-100)	0 (0-2)	0 (0-3)
fried	9	99 (97-100)	0 (0-2)	0 (0-2)
LAMB				
Best-end neck, cutlets, *raw*	13	45 (35-60)	28 (16-40)	26 (16-34)
barbecued	10	51 (34-61)	19 (11-34)	28 (15-40)
grilled	33	48 (30-73)	23 (5-49)	28 (16-40)
Breast, *raw*	10	66 (50-81)	33 (20-48)	0 (0)
roasted, medium	10	62 (46-86)	36 (13-54)	0 (0)
bone-in, raw	4	30 (29-32)	42 (35-48)	28 (23-36)
Chump chops, *raw*	26	67 (42-79)	21 (11-32)	12 (0 -26)
fried	21	65 (32-85)	20 (6-32)	14 (0-43)
Chump steaks, *raw*	12	80 (73-89)	21 (10-26)	0 (0)
fried	14	80 (57-94)	19 (5-41)	0 (0)
Leg chops, *grilled*	13	79 (53-92)	5 (2-10)	15 (5-45)
extra lean[b], *raw*	4	86 (80-92)	7 (4-12)	7 (4-11)

[a] Including fat tied on
[b] Extra lean is sold as Tenderlean in some supermarkets

	Number of samples	Means and ranges (%)		
		Trimmable lean	Trimmable fat	Inedible
Lamb continued				
Leg steaks, grilled	16	92 (84-98)	8 (2-15)	0 (0)
extra lean[a], raw	6	93 (84-99)	4 (1- 7)	3 (0-8)
Leg whole, raw	3	65 (59-71)	13 (8-17)	22 (20-24)
roasted, medium	10	67 (62-73)	8 (5-11)	23 (18-31)
-, well-done	10	67 (61-73)	8 (4-12)	24 (17-31)
New Zealand, roasted, medium	7	67 (53-85)	6 (2-12)	26 (15-38)
-, frozen, raw	10	64 (55-70)	12 (6-23)	23 (15-32)
-, -, roasted, medium	10	65 (60-70)	8 (3-12)	25 (19-29)
half fillet, braised	10	66 (43-81)	13 (4-31)	20 (11-29)
half knuckle, pot-roasted	10	65 (58-71)	8 (5-12)	26 (20-34)
extra lean[a], raw	4	66 (63-69)	12 (10-15)	22 (16-25)
boneless, roasted, medium	10	90 (84-93)	8 (5-14)	0 (0)
Loin chops, raw	17	56 (43-65)	22 (14-29)	21 (13-32)
grilled	33	61 (47-75)	19 (6-36)	19 (9-30)
microwaved	32	58 (27-72)	23 (9-42)	18 (10-31)
roasted	35	56 (27-74)	21 (6-37)	22 (12-38)
frozen, New Zealand, grilled	44	59 (39-74)	20 (4-35)	21 (9-42)
extra lean[a], raw	15	50 (31-61)	24 (19-41)	26 (18-37)
Loin joint, raw	4	50 (43-56)	26 (24-29)	23 (17-31)
roasted, medium	10	56 (48-68)	16 (8-27)	26 (19-34)
Neck fillet, raw	10	87 (75-98)	12 (1-24)	0 (0)
slices, grilled	10	89 (72-98)	9 (1-22)	0 (0)
strips, stir-fried	10	84 (69-96)	14 (2-30)	0 (0)
Rack of lamb, raw	4	50 (42-63)	22 (18-25)	28 (16-32)
roasted	10	48 (25-76)	25 (5-37)	25 (10-37)
Shoulder, whole, raw	3	60 (58-63)	19 (17-21)	20 (18-25)
roasted, medium	10	60 (48-65)	17 (12-23)	21 (18-28)
frozen, New Zealand, roasted, medium	9	58 (52-61)	19 (13-23)	22 (19-33)
half bladeside, pot-roasted	10	57 (52-66)	22 (15-27)	19 (7-29)
half knuckle, braised	10	57 (40-72)	20 (10-27)	22 (15-31)
extra lean[a], raw	4	58 (52-62)	24 (19-32)	18 (15-22)
boneless, roasted, medium	10	82 (70-97)	16 (2-27)	0 (0)
diced, kebabs, grilled	10	85 (62-100)	15 (0-38)	0 (0)
Stewing lamb, raw	10	53 (33-99)	13 (1-27)	33 (0-56)
pressure-cooked	10	55 (39-98)	12 (2-25)	33 (0-59)
stewed	10	56 (43-91)	10 (0-25)	32 (0-49)

[a] Extra lean is sold as Tenderlean in some supermarkets

	Number of samples	Means and ranges (%)		
		Trimmable lean	Trimmable fat	Inedible
PORK				
Belly, joint/slices, *raw*	12	60 (29-81)	32 (19-45)	8 (0-26)
joint, *roasted*	10	57 (46-72)	31 (16-48)	11 (0-29)
slices, *grilled*	24	49 (16-70)	32 (12-62)	15[a] (0-37)
Chump chops, *raw*	8	71 (50-90)	15 (7-23)	18 (3-27)
fried	6	62 (46-81)	22 (11-34)	15 (8-23)
Chump steaks, *raw*	13	91 (86-100)	9 (0-14)	0 (0)
fried	18	91 (80-100)	9 (0-21)	0 (0)
Diced, *raw*	10	91 (81-99)	8 (1-18)	0 (0)
casseroled	10	96 (90-100)	2 (0-6)	0 (0)
slow-cooked	10	98 (78-100)	2 (0-6)	0 (0)
kebabs, *grilled*	10	96 (88-100)	4 (0-12)	0 (0)
Fillet, *raw*	10	97 (94-100)	3 (0-6)	0 (0)
grilled	10	98 (94-100)	2 (0-6)	0 (0)
stir-fried	10	99 (93-100)	1 (0-7)	0 (0)
Hand, shoulder, boneless, *raw*	10	73 (59-86)	26 (14-40)	0 (0-1)
pressure-cooked	10	66 (59-79)	33 (20-40)	0 (0)
roasted	10	71 (55-83)	18 (16-43)	0[a] (0)
bone-in, *raw*	5	62 (46-72)	15 (14-20)	22 (9-39)
Leg, *raw*	10	68 (52-80)	18 (12-28)	13 (4-28)
knuckle/shank, *raw*	3	48 (35-64)	18 (16-21)	33 (16-44)
fillet, *raw*	2	74 (73-75)	15 (13-16)	12 (11-12)
Leg joint, *microwaved*	10	64 (49-78)	16 (7-27)	16[a] (9-35)
roasted, medium	10	69 (60-82)	14 (7-24)	15 (6-22)
-, well-done	10	67 (52-79)	17 (12-23)	14[a] (6-29)
frozen, roasted, medium	4	71 (66-75)	7 (6-7)	14[a] (11-21)
Loin chops, *raw*	6	59 (40-71)	25 (16-34)	16 (5-26)
barbecued	19	68 (49-83)	15 (8-21)	16 (0-29)
grilled	22	60 (44-74)	15 (7-27)	23[a] (13-39)
microwaved	18	65 (51-74)	14 (6-24)	20 (11-31)
roasted	16	59 (46-73)	17 (8-32)	24 (8-34)
frozen, grilled	34	55 (38-76)	15 (6-25)	29 (13-44)
Loin joint, *raw*	6	61 (53-66)	20 (12-25)	19 (15-25)
pot-roasted	10	55 (42-66)	19 (22-36)	15 (2-26)
roasted	10	58 (45-67)	16 (14-26)	21[a] (10-30)

[a] Excludes crackling. Crackling constitutes 3(0-13)% for belly slices, grilled; 9(5-15)% for hand-shoulder, roasted; 2(0-14)% for leg joint, microwaved; 1(0-8)% for leg joint, roasted, well-done; 7(3-11)% for leg joint, frozen, roasted, medium; 1(0-14)% for loin chops, grilled; 4(0-11)% loin joint, roasted

	Number of samples	Means and ranges (%)		
		Trimmable lean	Trimmable fat	Inedible
Pork *continued*				
Loin steaks, *raw*	24	80 (65-97)	20 (3-35)	0 (0)
fried	22	76 (54-89)	23 (10-45)	0 (0)
Spare-rib chops, *braised*	24	65 (40-100)	10 (0-30)	23 (0-47)
joint, *pot-roasted*	10	67 (52-90)	32 (8-44)	0 (0)
Spare-ribs, sliced, *raw*	10	41 (28-52)	6 (0-9)	52 (40-68)
grilled	10	39 (28-53)	6 (0-16)	54 (34-63)
Steaks, *raw*	15	89 (64-97)	11 (3-35)	0 (0)
grilled	19	92 (77-98)	8 (2-21)	0 (0)
stewed	15	89 (69-97)	9 (2-30)	0 (0)

PROPORTIONS OF LIGHT MEAT, DARK MEAT, SKIN AND BONE IN CHICKEN AND TURKEY

Every sample of raw and cooked chicken and turkey was separated before analysis into its light meat, dark meat, skin and bone where appropriate, and in the cooked samples the skin was further separated into crisp or dry skin (which contains little fat or water), moist skin and "under skin" (the skin under roasted birds where juices adhere and which is generally not eaten). The amounts were weighed, and the proportions are given below. It is important to realise that each set of values was derived from different samples. Small differences may therefore reflect differences between the samples as much as the effect of cooking.

Chicken and turkey are all with bone except where stated.

Whole chicken and turkey refers to the dressed carcase weight, without head, feet, feathers and giblets.

CHICKEN [a]

	Number of samples	Means and ranges (%)					
		White meat	Dark meat	Dry skin	Moist skin	Under skin	Bone
Raw							
Whole chicken, small	5	19 (15-24)	26 (23-28)	0	16 (13-19)	0	39 (33-41)
large	5	23 (20-25)	28 (26-30)	0	17 (13-22)	0	32 (31-33)
corn-fed	6	27 (19-31)	23 (20-29)	0	15 (13-18)	0	35 (34-37)
Poussin	5	20 (18-23)	23 (20-28)	0	15 (12-18)	0	42 (36-49)
Leg quarter	5	0	53 (49-58)	0	16 (12-23)	0	31 (28-36)
Wing quarter	5	43 (37-49)	7 (0-11)	0	16 (13-17)	0	34 (26-39)
Casseroled							
With skin							
Breast	1	81	0	0	9	0	10
Breast, without bone	3	89 (88-92)	0	0	10 (8-12)	0	0
Drumsticks	4	0	55 (52-59)	0	10 (8-14)	0	34 (26-38)
Leg quarter	4	13 (12-14)	39 (37-40)	0	11 (0-14)	0	37 (34-40)
Thigh	4	0	64 (63-67)	0	19 (17-21)	0	16 (14-20)
Wing quarter	4	40 (36-44)	14 (12-17)	0	13 (12-17)	0	33 (31-36)
Without skin							
Breast	1	71	0	0	0	0	29
Drumsticks	2	0	70 (62-79)	0	0	0	30 (2-39)
Leg quarter	2	19 (13-25)	37 (26-49)	0	0	0	44 (38-49)
Thigh	2	0	76 (72-79)	0	0	0	24 (21-27)
Wing quarter	2	56 (55-56)	17 (13-20)	0	0	0	27 (23-32)
Deep-fried							
Half	1	29	32	11	0	0	28
Leg quarter	5	1 (0-5)	54 (49-60)	12 (8-14)	0	0	33 (28-37)
Leg and wing	3	24 (19-31)	36 (31-39)	11 (9-12)	0	0	29 (26-32)
Wing quarter	1	40	10	20	0	0	31

a Chicken has been classified according to the following weights:
small/standard 1.1-1.4kg, medium 1.4-1.8kg, large 1.8-2.3kg, super 2.3-2.7kg, family 2.7+kg.

Chicken *continued*

	Number of samples	White meat	Dark meat	Dry skin	Moist skin	Under skin	Bone
					Mean and ranges (%)		
Grilled							
With skin							
Breast	10	72 (65-93)	0	6 (3-10)	2 (0-8)	0	20 (13-27)
Breast, without bone	7	91 (88-94)	0	6 (5-10)	1 (0-2)	0	0
Without skin							
Breast	2	91 (91,91)	0	0	0	0	9 (9,9)
Roasted							
Whole chicken, small	2	21 (19-23)	27 (27,27)	4 (4,4)	3 (2-4)	3 (3,3)	42 (39-45)
medium	2	21 (21,21)	28 (28,28)	5 (4-5)	5 (5,5)	5 (4-5)	37 (35-38)
large	1	28	25	4	3	3	38
super	3	28 (24-35)	29 (23-32)	6 (5-6)	3 (2-4)	2 (2-3)	32 (30-33)
family	2	30 (30,30)	27 (24-30)	5 (3-6)	4 (1-7)	1 (0-2)	33 (30-36)
frozen, small	3	25 (24-29)	30 (29-32)	5 (4-7)	4 (1-6)	1 (1-3)	33 (31-37)
medium	2	27 (25-28)	29 (28-30)	3 (1-5)	4 (2-6)	3 (2-3)	35 (33-37)
large	1	24	30	5	4	1	36
super	1	21	31	5	5	3	35
family	2	25 (25-26)	31 (30-32)	6 (5-6)	1 (1-2)	2 (2,2)	34 (35-33)
corn-fed	6	24 (23-27)	26 (23-30)	6 (4-8)	4 (2-5)	2 (2-3)	37 (35-42)
Drumsticks	20	0	57 (52-62)	4 (0-7)	3 (1-6)	0	37 (31-42)

	Number of samples	White meat	Dark meat	Dry skin	Moist skin	Under skin	Bone
				Mean and ranges (%)			
TURKEY [a]							
Raw							
Whole turkey, small	7	28 (22-35)	30 (23-39)	0	9 (6-10)	0	34 (27-40)
medium	7	31 (24-36)	28 (23-32)	0	8 (7-10)	0	34 (28-41)
extra large	7	37 (30-43)	26 (22-33)	0	9 (5-14)	0	27 (22-32)
Roasted							
Whole turkey, small	3	27 (25-29)	27 (25-28)	3 (2-4)	2 (1-5)	1 (1-2)	39 (38-40)
medium	2	31 (29-33)	28 (27-29)	3 (2-4)	1 (1,1)	2 (2,2)	35 (33-36)
large	3	38 (34-40)	26 (22-29)	4 (3-6)	2 (1-3)	2 (1-2)	29 (23-31)
extra large	2	38 (31-45)	27 (23-33)	4 (3-5)	2 (1-3)	1 (1-2)	28 (25-30)
frozen, medium	5	29 (26-32)	30 (23-34)	4 (2-5)	2 (2-3)	2 (1-2)	33 (29-38)
large	2	29 (26-33)	30 (30,30)	3 (2-3)	3 (3,3)	2 (1-2)	33 (31-36)
extra large	1	26	36	3	2	3	30
self-basting, frozen, small	1	26	23	2	3	2	43
medium	4	28 (26-29)	31 (29-33)	3 (2-4)	2 (1-3)	2 (1-2)	34 (32-35)
large	2	34 (28-40)	27 (24-30)	3 (2-3)	1 (1-2)	3 (1-4)	32 (30-34)
extra large	2	30 (29-32)	32 (30-34)	4 (3-5)	1 (0-2)	2 (1-2)	31 (31-32)
Drumsticks	9	0	66 (60-75)	4 (2-6)	2 (1-4)	0	27 (20-34)

[a] Turkey has been classified according to the following weights: small/mini 1.1-2.3kg, medium 2.3-4.5kg, large 4.5-7.3kg, extra large 7.3+kg.

THE COMPOSITION OF EXTRA LEAN MEAT

Butchery techniques which carefully separate the muscles and remove all possible fat from between them are known as seam butchery, and are of increasing importance. Cuts produced from seam butchery differ from traditional cuts, are particularly low in fat, and are now available at retail and increasingly found on restaurant menus. The amounts of the main nutrients in representative samples of these cuts are presented in this appendix.

Further details can be obtained from the references in this appendix, and from the Meat and Livestock Commission, Milton Keynes MK6 1AX.

Composition of extra lean meat per 100g

Food		Edible Proportion g	Water g	Total Nitrogen g	Protein g	Fat g	Carbo-hydrate g	Energy value kcal	Energy value kJ
Beef									
Lean cubes/stir fry	43 samples	1.00	73.1	3.58	22.4	3.4	0	120	503
Mini joints/steak bars	43 samples	1.00	74.3	3.42	21.4	3.4	0	116	486
Skirt steaks	43 samples	1.00	70.9	3.42	21.4	6.9	0	148	618
Pork									
Escalopes/nuggets	32 samples	1.00	75.8	3.46	21.6	1.7	0	102	426
Medallions	32 samples	1.00	73.1	3.54	22.1	3.9	0	124	517
Mini leg joints	32 samples	1.00	75.8	3.46	21.6	1.7	0	102	426
Lamb									
Leg steaks/stir fry/mini joints	27 samples	1.00	74.1	3.23	20.2	5.2	0	128	534
Medallions	27 samples	1.00	70.0	3.31	20.7	8.0	0	152	648

Composition of extra lean meat per 100g

Food	Starch g	Total sugars g	Dietary fibre Southgate method g	Dietary fibre Englyst method g	Satd g	Fatty acids cis & trans Mono-unsatd g	Fatty acids cis & trans Poly-unsatd g	Total Trans g	Cholesterol mg
Beef									
Lean cubes/stir fry	0	0	0	0	1.37	1.44	0.16	N	N
Mini joints/steak bars	0	0	0	0	1.38	1.45	0.15	N	N
Skirt steaks	0	0	0	0	2.90	2.86	0.25	N	N
Pork									
Escalopes/nuggets	0	0	0	0	0.62	0.62	0.27	N	N
Medallions	0	0	0	0	1.52	1.44	0.54	N	N
Mini leg joints	0	0	0	0	0.62	0.62	0.27	N	N
Lamb									
Leg steaks/stir fry/mini joints	0	0	0	0	N	N	N	N	N
Medallions	0	0	0	0	N	N	N	N	N

REFERENCES

Analytical Methods Committee (1991) Nitrogen factors for pork: a reassessment. *Analyst* **116**, 761-766

Analytical Methods Committee (1993) Nitrogen factors for beef: a reassessment. *Analyst* **118**, 1217-1226

Analytical Methods Committee (1995) Nitrogen factors for sheepmeat: Part 2. *Analyst*, In press

COOKING METHODS

All cooking water was unsalted and the fat used in frying was corn oil. Cooking liquids and any juices produced during cooking were discarded before analysis. All times given are total cooking times.

Barbecuing

Beef	– fillet, rump, sirloin steaks	– rare 5–10 mins
		– medium 9–13 mins
		– well-done 13–19 mins
Lamb	– best end neck cutlets	8–12 mins
Pork	– loin chops	16–20 mins

Boiling

Beef	– brisket, silverside	25 mins per 450g plus 25 mins extra
Chicken	– giblets	45 mins
Turkey	– giblets	90 mins
Brain	– lamb	20 mins after soaking for 2 hours
Trotters & tails		2 hours

Braising

Brown in 1 tablesp oil per 450g, cover with a small amount of water and cook in preheated oven 170°C to 180°C/Mark 3 to 4.

Beef	– braising steak	25–30 mins per 450g plus 30 mins
Lamb	– leg and shoulder joint	30 mins per 450g
Pork	– spare-rib chops	1–1½ hours

Casseroling

Brown in 1 tablesp oil, cover with 600 ml per 450g or per chicken and cook in preheated oven 170°C to 180°C/Mark 3 to 4.

Chicken or turkey	– diced	¾ – 1 hour
Chicken	– portions	1–1¼ hours
Pork	– diced	1–1½ hours

Frying

Cook with 1 tablesp oil per 450g.

Beef	– fillet, rump, sirloin steaks	– rare 5–10 mins – medium 9–13 mins – well-done 13–19 mins
Veal	– escalope	10 mins
Lamb	– chump chops/steaks	12–16 mins
Pork	– chump chops – loin steaks	16–20 mins – 1–2cm thick 12–16 mins – over 2cm thick 16–20 mins
Kidney	– lamb, pig	– 12–15 mins
Liver	– calf, chicken, lamb	12–15 mins
Sweetbread	– lamb	5–8 mins after soaking for 2 hours and boiling for ½ hour

Grilling

Cook in a preheated moderate grill. For kebabs, thread 2.5 cm cubes onto skewers.

Beef	– fillet, rump, sirloin steaks	– rare 5–10 mins – medium 9–13 mins – well-done 13–19 mins
Lamb	– best end cutlets – kebabs – leg chops/steaks – loin chops – neck fillet slices	8–12 mins 10–15 mins – 1–2cm thick 8–12 mins – over 2cm thick 12–16 mins 12–16 mins 6–10 mins
Pork	– fillet slices – kebabs – loin chops – spare ribs – steaks	6–10 mins 15 mins 16–20 mins 20–30 mins – 1–2cm thick 12–16 mins – over 2cm thick 16–20 mins
Chicken	– breast	– skinless 15 mins – skin on 20 mins
Turkey	– breast	15–20 mins

Microwaving

Turn meat half way through cooking.

Beef	– fore-rib	Medium power, 650W oven for 40 mins; stand for 15 mins
	– mince	Medium power, 650W oven for 4–8 mins
	– topside	High power, 650W oven for 7–8 mins per 450g; stand for 15–20 minutes or to centre temperature of 70°C
Lamb	– loin chops	High power, 800W oven; stand 3 mins – 225g 3–4 mins – 450g 6–7 mins
Pork	– leg joint	High power, 9 mins per 450g; stand 20–30 mins
	– loin chops	High power, 800W oven; stand 3 mins – 150g 2 mins – 350g 3–4 mins – 450g 4–5 mins

Pot roasting

Brown in 1 tablesp oil, add 150ml water per joint, bring to the boil, cover, reduce heat and simmer on hob.

Beef	– flank, silverside	30–40 mins per 450g plus 30–40 mins extra
Lamb	– leg and shoulder joints	25–30 mins per 450g plus 25–30 mins extra
Pork	– loin joint, spare-rib chops	30–35 mins per 450g plus 30–35 mins extra

Pressure cooking

Cook at low pressure (15 psi).

Beef	– stewing steak	Per 450g: brown in 1 tablsp oil, add 300 ml water and cook for 20 mins
Lamb	– stewing lamb	300ml water per 675–900g meat, 15 mins per 450g.
Pork	– shoulder joint	Brown in 1 tablesp oil, add 600 ml water, cook 15 mins per 450g

Roasting

Cook uncovered in preheated oven unless otherwise specified.

Beef	– fore-rib, sirloin, topside	15 mins at 220˚C/Mark 7, then lower to 180˚C/Mark 4 – medium 25 mins per 450g plus 25 mins extra to centre temperature 70˚C – well-done 30 mins per 450g plus 30 mins extra to centre temperature of 80˚C
Lamb	– rack of lamb, breast, leg, loin and shoulder joints	180˚C/Mark 4 – medium 25 mins per 450g plus 25 mins extra to centre temperature 70–75˚C – well-done 30 mins per 450g plus 30 mins extra to centre temperature of 75–80˚C
	– loin chops	Covered at 180˚C/Mark 4 for 40–60 mins for 4 chops. Uncover for the last 5–10 mins
Pork	– belly and loin joints	180˚C/Mark 4 for 30 mins per 450g plus 30 mins extra to centre temperature 75–80˚C.
	– leg joint	180˚C/Mark 4. – medium 30 mins per 450g plus 30 mins extra to centre temperature 75–80˚C – well-done 35 mins per 450g plus 35 mins extra to centre temperature 80–85˚C
Chicken	– drumsticks	25 mins at 190˚C/Mark 5
	– whole	Covered for 20 mins per 450g plus 20 mins extra at 190˚C/Mark 5. Uncover for the last 15 mins
Turkey	– drumsticks	35–45 mins at 190˚C/Mark 5
	– whole	Covered at 190˚C/Mark 5; uncover for last 40 mins. Up to 7.5 kg, 20 mins per 450g plus 20 mins extra. Over 7.5 kg, 15 mins per 450g plus 15 mins extra
Heart	– lamb	90 mins at 200˚C/Mark 6

Slow cooking

Beef	– braising steak	Brown in 1 tablesp oil per 450g, add small amount of water, bring to boil. Transfer to a slow cooker and cook for 7 hours on low power
Pork	– diced pork	Brown in 1 tablesp oil per 675g, add 450ml water, bring to boil. Transfer to a slow cooker and cook for 7–9 hours on low power

Stewing

Mince: unless otherwise specified, dry fry for 4–6 minutes per 225–450g, drain fat off, add water, cover and simmer on hob. Other cuts: brown in 1 tablesp oil, add water, cover and simmer on hob

Beef	– mince	300ml water per 450g; cook for 1 hour
	– stewing steak	600ml water per 450g; cook for 2 hours
Veal	– mince	300ml water per 450g; cook for 1 hour
Lamb	– mince	150ml water per 450g; cook for 30 mins
	– stewing lamb	600ml water per 450g; cook for 1–1½ hours
Pork	– mince	150 ml water per 450g; cook for 30 mins
	– steaks	600 ml water for 4 steaks; cook for 1–1½ hours
Turkey	– mince	Brown in 1 tablesp oil; add 300 ml water per 450g; cook for 15–20 mins
Heart	– pig	90 mins at 200°C/Mark 6
Tongue	– ox	3¾ hours after soaking for 18 hours

Stir frying

Cook thin strips of meat in 1 tablesp oil over moderate heat.

Beef	– rump steak	5–10 mins per 450g
Lamb	– neck fillet	2–4 mins per 225g
Pork	– fillet	4–6 mins per 225g
Chicken or turkey	– breast strips	5 mins

WEIGHT LOSSES ON COOKING MEAT

When meat is cooked it loses fat and juices, and these were discarded. The weight of every sample that was cooked was recorded both before and after cooking, and the losses are given below. Small differences may reflect differences between the samples and the exact cooking conditions as much as the cooking methods themseleves.

Poultry included bones and skin expect where stated.

The proportions of bone and any other inedible material in the raw and cooked meats, as well as the amounts of trimmable fat, lean and skin where appropriate, are given in the Appendix on page 000.

	Number of samples	% loss Means and ranges
BEEF		
Braising steak, *braised*	10	40 (36-46)
slow-cooked	10	36 (25-43)
Brisket, *boiled*	10	34 (26-41)
Fillet steak, *grilled*	20	26 (15-42)
fried	16	24 (14-38)
Flank, *pot-roasted*	6	35 (28-42)
Fore-rib, *microwaved*	10	40 (34-46)
roasted	10	34 (28-41)
Mince, *barbecued*	10	35 (27-45)
microwaved	10	28 (19-43)
stewed	10	18 (4-25)
extra lean, *stewed*	10	18 (9-29)
frozen, *stewed*	10	7 (-7-17)[a]
Rump steak, *barbecued*	10	31 (18-47)
fried	10	27 (20-36)
grilled	10	28 (18-35)
stir-fried	10	29 (17-40)
Silverside, *pot-roasted*	10	39 (36-44)
salted, *boiled*	8	39 (29-50)
Sirloin joint, *roasted*	10	32 (25-39)
Sirloin steak, *fried*	10	22 (13-34)
grilled, rare	10	12 (6-26)
-, medium-rare	19	21 (7-40)
-, well-done	10	38 (27-59)
Stewing steak, *pressure-cooked*	10	40 (22-48)
stewed	10	36 (26-47)

[a] Two samples gained weight, as the juices formed a gel on cooling.

	Number of samples	% loss Means and ranges
Beef *continued*		
Topside, *microwaved*	10	38 (18-46)
roasted, medium-rare	10	32 (19-42)
-, *well-done*	10	42 (34-52)
VEAL		
Escalope, *fried*	9	38 (30-43)
Mince, *stewed*	5	28 (28-31)
LAMB		
Best-end neck, cutlets, *barbecued*	10	29 (23-38)
grilled	33	32 (15-54)
Breast, boneless, *roasted, medium*	10	28 (20-41)
Chump chops, *fried*	21	26 (8-45)
Chump steaks, *fried*	14	25 (12-36)
Leg chops, *grilled*	13	32 (23-45)
Leg steaks, *grilled*	16	32 (24-47)
Leg, whole, *roasted, medium*	10	31 (20-35)
roasted, well-done	10	35 (31-41)
New Zealand, fresh, *roasted, medium*	7	31 (30-32)
-, *frozen, roasted, medium*	10	32 (29-35)
half fillet, *braised*	10	25 (17-33)
half knuckle, *pot-roasted*	10	27 (14-33)
boneless, *roasted, medium*	10	36 (33-40)
Loin chops, *grilled*	33	31 (15-52)
microwaved	10	33 (24-41)
roasted	35	37 (17-57)
New Zealand, frozen, *grilled*	44	36 (23-49)
Loin joint, *roasted, medium*	10	26 (14-35)
Mince, *stewed*	10	28 (22-33)
frozen, *stewed*	8	26 (16-37)
Neck fillet, slices, *grilled*	10	37 (24-49)
strips, *stir-fried*	10	22 (16-35)
Rack of lamb, *roasted, medium*	10	25 (13-40)
Shoulder, whole, *roasted, medium*	10	32 (21-40)
whole, New Zealand, frozen, *roasted, medium*	9	32 (26-40)
half bladeside, *pot-roasted*	10	27 (17-33)
half knuckle, *braised*	10	19 (12-29)
boneless, *roasted, medium*	10	36 (30-41)
diced, kebabs, *grilled*	10	38 (24-48)
Stewing lamb, *pressure-cooked*	10	28 (22-40)
stewed	10	27 (21-41)

	Number of samples	% loss Means and ranges
PORK		
Belly, joint, *roasted*	10	29 (27-34)
slices, *grilled*	24	36 (29-61)
Chump chops, *fried*	6	20 (16-23)
Chump steaks, *fried*	18	27 (15-35)
Diced, *casseroled*	10	37 (31-41)
slow-cooked	10	34 (22-39)
kebabs, *grilled*	10	39 (32-47)
Fillet, *grilled*	10	37 (27-44)
stir-fried	10	33 (26-40)
Hand, shoulder, *pressure-cooked*	10	30 (23-38)
roasted	10	37 (30-43)
Leg joint, *microwaved*	10	36 (28-44)
roasted, medium	10	35 (23-41)
-, well-done	10	39 (35-44)
frozen, roasted, medium	4	37 (33-39)
Loin chops, *barbecued*	19	28 (15-48)
grilled	22	32 (21-40)
microwaved	18	32 (20-40)
roasted	16	38 (24-57)
frozen, grilled	34	36 (25-46)
Loin joint, *pot-roasted*	10	28 (11-34)
roasted	10	31 (26-36)
Loin steaks, *fried*	22	28 (21-38)
Mince, *stewed*	10	22 (17-28)
Pork steaks, *grilled*	19	38 (28-47)
stewed	15	39 (31-45)
Spare-rib chops, *braised*	26	32 (23-47)
Spare-rib joint, *pot-roasted*	10	31 (15-37)
Spare-ribs, sliced, *grilled*	10	29 (24-35)
CHICKEN		
Casseroled		
Breast	1	25
Breast, *casseroled without bone*	3	31 (27-37)
Breast, *casseroled without skin*	1	21
Breast, *casseroled without skin or bone*	1	25
Drumsticks	4	16 (11-18)
Drumsticks, *casseroled without skin*	2	23 (21-24)
Thighs	4	30 (28-33)
Thighs, *casseroled without skin*	2	26 (25-26)
Thighs, diced	10	35 (32-38)
Leg quarter	4	23 (14-28)
Leg quarter, *casseroled without skin*	2	23 (23,23)
Wing quarter	2	21 (18-24)
Wing quarter, *casseroled without skin*	1	24

	Number of samples	% loss Means and ranges
Chicken *continued*		
Grilled		
Breast	9	27 (18-31)
Breast, *grilled without bone*	7	29 (18-36)
Breast, *grilled without skin*	2	25 (25,25)
Breast, *grilled without skin or bone*	8	26 (21-33)
Stir-fried		
Breast strips	10	21 (17-29)
Roasted [a]		
Whole, fresh		
small	2	24 (22-26)
medium	2	14 (13-15)
large	1	26
super	3	37 (28-41)
family	2	26 (24-27)
Whole, frozen		
small	3	32 (28-37)
medium	2	29 (27-30)
large	1	24
super	1	30
family	2	29 (25-33)
Whole, corn fed, average	6	21 (10-29)
Drumsticks, fresh	10	23 (14-34)
frozen	10	29 (18-50)
TURKEY		
Breast fillet, *grilled*	9	32 (23-46)
Breast strips, *stir-fried*	8	23 (18-27)
Mince, *stewed*	5	33 (27-38)
Thighs, diced, *casseroled*	8	34 (27-38)
Roasted [b]		
Whole, fresh		
small	3	23 (21-25)
medium	2	26 (24-29)
large	3	27 (24-29)
extra large	2	29 (29-30)

[a] Chicken has been classified according to the following weights:
small/standard 1.1-1.4 kg, medium 1.4-1.8 kg, large 1.8-2.3 kg, super 2.3-2.7 kg, family 2.7+ kg

[b] Turkey has been classified according to the following weights: small 1.4-2.3 kg, medium 2.3-4.5 kg, large 4.5-7.3 kg, extra large 7.3+ kg

Number of	% loss samples	Means and ranges
Turkey *continued*		
Whole, frozen		
medium	5	28 (22-38)
large	2	29 (23-35)
extra large	1	35
Whole, self-basting		
small	1	31
medium	4	29 (20-39)
large	2	37 (35-38)
extra large	2	33 (27-39)
Drumsticks	8	26 (21-33)

INDIVIDUAL FATTY ACIDS

New analytical values for proportions of fatty acids in typical cuts of meat, poultry and game are given here. Most of the values, with the exception of some of the poultry and feathered game, are from analysis of raw meat. However, since there are only small differences in the proportions of fatty acids in raw and cooked meat, the values shown below can be applied to a wide range of the foods in the main tables. The amounts are presented *per 100 grams of fatty acids*, and can be converted to amounts per 100 g of fat in each item by dividing by the factors in the following table:

Conversion factors to give total fatty acids in fat

Beef lean	0.916	Poultry	0.945
Beef fat	0.953	Heart	0.789
Lamb lean	0.916	Kidney	0.747
Lamb fat	0.953		
Pork lean	0.910		
Pork fat	0.953		

The amounts of the individual fatty acids in different cuts of meat can be calculated from the proportions of lean and fat, and of white and dark meat (and skin) given in the Appendix on page 000.

Names of the fatty acids occurring in the tables:

No of carbon atoms and double bonds	Systematic name	Common name
Saturated acids		
12:0	Dodecanoic acid	Lauric acid
14:0	Tetradecanoic acid	Myristic acid
15:0	Pentadecanoic acid	
16:0	Hexadecanoic acid	Palmitic acid
17:0	Heptadecanoic acid	Margaric acid
18:0	Octadecanoic acid	Stearic acid
22:0	Docosanoic acid	Behenic acid
24:0	Tetracosanoic acid	Lignoceric acid

No of carbon atoms and double bonds	Systematic name	Common name
Monounsaturated acids		
14:1	Tetradecenoic acid	Myristoleic acid
16:1	Hexadecenoic acid	Palmitoleic acid
17:1	Heptadecenoic acid	
18:1 (*cis*)	Octadecenoic acid	Oleic acid
		cis-Vaccenic acid
18:1 (*trans*)		Elaidic acid
		trans-Vaccenic acid
20:1	Eicosenoic acid	Eicosenic acid
		Gadoleic acid
22:1	Docosenoic acid	Erucic acid
24:1	Tetracosenoic acid	Nervonic acid
		Selacholeic acid
Polyunsaturated acids		
18:2	Octadecadienoic acid	Linoleic acid
18:3	Octadecatrienoic acid	Linolenic acid
20:3	Eicosatrienoic acid	
20:4	Eicosatetraenoic acid	Arachidonic acid
20:5	Eicosapentaenoic acid	
22:5	Docosapentaenoic acid	Clupanodonic acid
22:6	Docosahexaenoic acid	Cervonic acid

Fatty acids, g per 100g fatty acids

	Saturated					Monounsaturated					Polyunsaturated		
	14:0	15:0	16:0	17:0	18:0	14:1	16:1	17:1	18:1	20:1	18:2	18:3	20:4
Beef, trimmed lean	2.8	0.5	24.8	1.1	14.5	0.7	4.1	1.0	41.8	0.3	3.3	0	0.5
Beef, trimmed fat	3.3	0.7	25.9	1.2	15.4	0.9	4.4	1.0	40.5	0.3	2.6	0	0
Beef, mince	3.1	1.2	24.9	2.4	16.4	0.8	4.1	1.0	41.8	0.2	2.3	0	0
Veal, escalope[a]	3.6	0.3	22.4	1.1	11.9	0.8	4.5	0.4	37.7	0.3	10.8	0	3.1
Lamb, trimmed lean[b]	5.3	0.7	21.8	1.1	17.6	0.2	2.2	0.7	38.4	0.3	2.7	2.0	0.3
Lamb, trimmed fat[b]	6.0	0.8	21.8	1.3	19.9	0.2	2.2	0.6	36.4	0.4	2.3	1.8	0.1
Pork, trimmed lean	1.2	0.1	23.1	0.3	12.5	0.1	2.5	0.3	38.2	0.9	14.9	1.3	1.1
Pork, trimmed fat	1.2	0.1	23.0	0.4	12.7	0.2	2.4	0.3	40.1	1.0	14.7	1.6	0.2
Chicken, skin	1.0	0.2	21.5	0.5	6.1	0.2	5.6	0.2	44.9	0.6	14.6	2.5	Tr
Chicken, dark meat	0.9	0.2	20.6	0.5	6.1	0.2	4.8	0.3	43.3	0.6	16.7	2.6	0.6
Chicken, light meat[c]	0.8	0.2	21.5	0.5	6.6	0.1	4.2	0.2	41.6	0.6	15.9	2.2	1.0
Turkey, dark meat[d]	1.3	0.4	20.6	0.7	9.0	0.2	3.8	0.3	33.8	0.7	21.2	2.5	1.2
Turkey, light meat[e]	1.0	0.3	22.0	0.7	10.3	0.1	3.5	0.3	32.4	0.6	18.3	1.5	1.6
Duck, meat, fat and skin	0	0	22.7	0.3	6.5	0	4.0	0.2	48.9	0.8	14.2	1.7	0
Grouse, meat only	0.6	0.1	16.7	1.5	5.7	0	1.9	0.3	10.7	0.1	31.9	30.3	Tr
Pheasant, meat only[f]	0.5	0	24.0	0.3	10.7	0	5.2	0	41.0	0.3	13.3	0	0
Rabbit, meat only[g]	2.2	0.5	28.7	1.0	8.4	0	3.6	0.9	19.6	0.2	10.5	23.2	0
Venison[h]	5.0	1.8	24.6	1.1	17.1	0	4.0	1.2	19.5	0.2	11.5	3.8	4.2
Heart, pig	1.2	0	25.2	0.5	19.6	0	1.9	0.8	31.8	0.8	14.0	0	2.3
Kidney, lamb[i]	0.6	0.2	19.8	1.3	20.7	0	2.3	1.1	24.2	1.2	9.9	2.5	6.9

[a]Contains 0.6g 20:3 and 0.8g 22:5 per 100g
[b]Contains 0.6g 12:0 per 100g
[c]Contains 0.6g 22:5 and 0.7g 22:6 per 100g
[d]Contains 0.7g 22:6 per 100g
[e]Contains 0.6g 22:1, 0.5g 22:5 and 0.9g 22:6 per 100g
[f] Contains 3.1g 22:1 per 100g
[g]Contains 1.4g 20:5 and 3.3g 22:5 per 100g
[h]Contains 1.9g 22:0, 1.8g 24:0, 1.0g 24:1 and 4.3g 20:5 per 100g
[i]Contains 0.7g 22:1 per 100g

VITAMIN A FRACTIONS AND RANGES IN OFFAL

Carcase meat, poultry and game contain little vitamin A, but offals can contain much more, with 13-*cis* retinol being present as well as the *trans* form. Where the amounts of these fractions have been determined, they are shown below. The total retinol equivalent has been taken as the sum of the *trans*-retinol plus 75% of the 13-*cis*-retinol (Sivell *et al.*, 1984)[a].

Retinol fractions, μg per 100g food

No. 18-	Food	All-*trans*-retinol	13-*cis*-retinol	Retin-aldehyde	Dehydro-retinol	Retinol Equivalent
393	**Giblets, chicken**, *raw*	5280	315	0	0	5515
395	**Giblets, turkey**, *cooked*	2780	465	0	0	3130
402	**Kidney, lamb**, *raw*	85	15	0	0	96
404	**Kidney, ox**, *raw*	90	20	0	0	105
406	**Kidney, pig**, *raw*	100	20	0	0	115

The amounts of vitamin A in liver are very variable, as shown in the second table. The amounts depend not only on the levels in the grass and feed, but also the age of the animal.

Ranges of total retinol in liver, μg per 100g food

No. 18-	Food	Number of samples	Mean[b]	Standard deviation[b]	Range[c]
409	**Liver, calf**, *raw*	42	18,800	12,500	2,100-40,700
411	**Liver, chicken**, *raw*	125	9,700	4,400	2,500-18,800
413	**Liver, lamb**, *raw*	228	17,300	10,400	3,300-52,500
415	**Liver, ox**, *raw*	121	14,200	11,000	500-39,500
417	**Liver, pig**, *raw*	133	17,400	11,800	4,200-35,000

[a] Sivell, L.M., Bull, N.L., Buss, D.H., Wiggins, R.A., Scuffam, D., Jackson, P.A. (1984). Vitamin A activity in foods of animal origin. *J. Sci. Fd Agric.* **35**, 931-939

[b] Means and standard deviations calculated from all values

[c] Excluding the two highest and the two lowest values

VITAMIN D FRACTIONS

Meat can contain vitamin D (cholecalciferol) derived from the action of sunlight or, for pigs and poultry, from the feed. Circulating vitamin D is mainly in the form of the more active 25-hydroxy vitamin D. The amounts of both forms were analysed by HPLC in a range of meats, and the values are shown below.

The amounts of vitamin D shown in the tables have been interpolated from these values, with total vitamin D activity taken as the sum of the cholecalciferol and five times the amount of 25-hydroxycholecalciferol.

Vitamin D fractions, μg per 100g food

| No. 18- | Food | Vitamin D fractions | | Total Vitamin D |
		Vitamin D_3	25-hydroxy Vitamin D_3	
Beef				
1	Beef, lean	Tr	0.08-0.16	0.5
3, 5	Beef, fat	Tr	(Tr)	(Tr)
38	Mince, *stewed*	Tr	0.18	0.9
70	Sirloin steak, *grilled, medium-rare*, lean	Tr	0.09	0.5
76	Stewing steak, *raw*, lean	Tr	0.16	0.8
80	Stewing steak, *stewed*, lean	Tr	0.14	0.7
88	Topside, *roasted, medium-rare*, lean	Tr	0.08	0.4
Veal				
93	Escalope, *fried*	1.2	0.03	1.4
Lamb				
96	Lamb, lean	Tr-0.5	0.02-0.07	0.4
98, 100	Lamb, fat	Tr-0.4	Tr-0.1	0.5
111	Breast, *raw*, lean	Tr	0.07	0.4
135	Leg, whole, *roasted, medium*, lean	0.5	0.05	0.7
159	Mince, *stewed*	0.1	0.08	0.5
179	Shoulder, whole, *roasted*, lean	0.5	0.05	0.8
199	New Zealand lamb shoulder, *roasted*, lean	0.5	0.03	0.7

Vitamin D fractions, μg per 100g food

No. 18-	Food	Vitamin D fractions Vitamin D$_3$	25-hydroxy Vitamin D$_3$	Total Vitamin D
Pork				
201	Pork, lean	Tr-0.4	Tr-0.12	0.5
203, 205	Pork, fat	1.1	0.05	1.3
208	Belly joint, *roasted*, lean and fat	0.6	0.08	1.0
233	Hand, shoulder joint, *roasted*, lean	0.3	0.07	0.7
240	Leg joint, *roasted medium*, lean	Tr	0.12	0.6
251	Loin chops, *grilled* lean	0.3	0.09	0.8
271	Spare rib chops, *braised*, lean	0.4	Tr	0.4
Chicken				
292	Skin, *raw*	0.6	0.26	1.9
299	Corn-fed, dark meat, *raw*	Tr	0.09	0.5
300	Corn-fed, light meat, *raw*	Tr	0.09	0.5
329	Dark meat, *roasted*	Tr	Tr	0.1
330	Light meat, *roasted*	Tr	0.10	0.5
332	Dry skin, *roasted*	0.7	0.05	1.0
333	Moist skin, *roasted/ grilled*	0.1	0.09	0.6
Turkey				
351	Skin, *raw*	0.6	0.03	0.8
358	Dark meat, *roasted*	0.3	Tr	0.3
359	Light meat, *roasted*	0.1	Tr	0.1

FOOD INDEX

Foods are indexed by their food number and **not** by their page number, except for foods appearing in the Extra Lean Cuts Appendix on page 136.

A number of entries refer to more common names for, or the likely origin of, the specified cut of meat. A list of these names is given in the Alternative and Taxonomic Names Appendix on page 120.